Keepers of the Land

A Celebration of Canadian Farmers

Keepers of the Land

A Celebration of Canadian Farmers

CARL HIEBERT & DEB CRIPPS

GIFT OF WINGS PUBLISHING

Published by
Gift of Wings Publishing
60 Adelaide St.
Linwood, ON.
N0B 2A0

carl@giftofwings.ca

Library and Archives Canada Cataloguing in Publication

to follow

Design by Gillian Stead

Printed in Canada by Friesens Printers, Altona, Manitoba

Photographing Ken Farkash's buffalo, Vermilion, Alberta. Riding on quads (all terrain vehicles) was a huge help in accessing photo opportunities.

Introduction

June 1, 2006. Leaving the Pacific, we begin our cross-country adventure.

THE MOST UNLIKELY of cross-Canada ventures started with a boyhood memory. When I was about 12, Dad finally agreed I was old enough to run the "big" tractor. I could spend a day transforming a fading green pasture into straight dark rows. Ploughing. No other job on our 67-acre farm could have been more satisfying.

Sometimes, the field seemed to stretch on forever, and time moved even more slowly than my tractor creeping along in third gear. My stomach growled but the last peanut butter sandwich had disappeared an hour ago. Yet the day was rich in smells and activity and had a feeling that all was right in the world.

Wheeling, screaming flocks of seagulls all vied to be closest to the fresh-turned earth, scrambling for a lunch of earthworms and grubs. Our dog King kept well ahead, leading me up and down the field and then occasionally charging the gulls to break the boredom. The plough shears cleanly sliced, folded, reshaped the land. The smell of fresh earth rose around me…subtle, almost musty, ultimately distinctive.

After a while, you become a part of the machine. The throttle, gearshift and spinner fit intuitively into your palm. Inevitably, the engine's roar mesmerizes, steady and undaunted,

lugging only occasionally as we hit a patch of hard clay. My neighbour buddies bragged about their father's Massy Harris or John Deere tractors but in my estimation, nothing could touch my dad's W4 McCormick.

I left that tractor and farm behind when I was 18. Farm life seemed hard and I hoped to carve an easier path. That path eventually led me to become a traveller, salesman, pilot, photographer, author, and presenter. But as the years and miles between me and the farm grew, so did the memories. Time gave me a new appreciation for the importance of food and its production. A new-found respect for the tenacity and commitment of farmers, for the challenges they face. These are the unsung heroes of our country, I realized, and their story needs to be told.

And in one of those quiet but undeniable "ah-ha" moments, I knew what had to be done. I would publish a book to celebrate these people, with the book proceeds given to charity. I would cross Canada meeting as many farmers as I could so I could share their stories. And I'd do it on a W4 McCormick like my dad's. Every time I thought of this I wanted to smile. Once again I would be a young boy, carving another black furrow across the field.

This is how *Keepers of the Land* came to be.

Arguably, Ol' Red was not the fastest vehicle on the highway.

Embarking on a 6,000 kilometre cross-Canada trek on a 57 year-old tractor was not without surprises and challenges. First I needed to modify my unlikely steed. My 1949 W4 was never designed for such a trip — nor was I. My farmer friend Ernest and I spent hours fabricating hand controls for the clutch and brake. At times I'd be driving in city traffic in high gear but would need immediate and precise control. How could I simultaneously and safely manage four controls — steering wheel, throttle, clutch and brake — with only two hands? Practice, I concluded. Lots of practice.

The old dish-pan, not-so-forgiving steel seat had to go. We replaced it with an air-ride model with adequate back support. A rack mounted behind me would carry my wheelchair. Then there was the matter of creature comforts. I had initially hoped to keep the tractor as close to its original form as possible. No air-conditioned, climate-controlled cab for this adventurer. But farmers, wiser and more weathered, convinced me otherwise. I would at least need a protecting roof. Engine and mileage tests, new spark plugs and new tires and I was finally ready for the road.

June 1, 6 a.m., Whiterock Beach, B.C. It's a warm enough day but a pale overcast and a sense of foreboding nags at my public optimism. I start at the very edge of the country, carefully backing Ol' Red into the Pacific almost half way up the rear tires. I didn't need CBC live national coverage showing that I should have considered a boat. The camera rolls, I ease out the clutch and begin my eastward-bound adventure.

Within an hour, it begins to rain. And rain. For the first three weeks of the trip, well into Manitoba, we will experience only four days of sunshine. Quickly enough, I realize the shortcomings in some of my plans. While I'm cruising full throttle in road gear at 28 kph, the rain seems intent on making my ride one of misery. The smaller front tires spin much faster than the rear ones, blasting two fire-hose-directed-showers against my body. I adjust goggles over my glasses and hang a temporary Plexiglas windshield in front of me, imagining the problem is fixed. Good, except I haven't counted on heavy traffic forcing me off the rain swept and relatively clean highway and over on the shoulder. Within seconds, the splattering debris transforms my windshield into an opaque, brown mudshield. Until I can get back on the highway in the fresh clean rain, my only choice is to lean over and look out either side. Days later, helpful Hutterites near Saskatoon put their innovative manufacturing skills into gear and outfit Ol' Red with a much appreciated set of front fenders. One problem finally solved.

Despite the rain and heavy overcast, the ride through the Rockies is breathtaking and invigorating. As we chug up the final part of the ever-increasing incline of the Coquihalla Pass, Ol' Red almost meets her match. Blue smoke spews upwards and the engine temperature gauge heads toward the red danger line. As the RPM's drop ever lower, I try to will her with intent. Never have schoolboy memories of my grade-one reader seemed more apt: "I think I can, I think I can." Shifting to the next-lower gear means slowing us to not much more than walking speed and adding several hours to the day's ride. For the next few kilometres, we ride on the verge of only barely making it. Just when the effort seems greatest, the sign welcoming us to the summit finally appears. And Ol' Red and I now have a new understanding of each other. We can do this! We will do this!

The Rockies are undeniably awe-inspiring but just as during my two cross-Canada ultralight flights, I welcome the prairies. These will be days of endless highways with wide shoulders, sweeping vistas and easy miles.

The perfect morning dawns on the day we leave North Battleford, heading for Waldheim, Saskatchewan. The anticipation of the day is enough to override my normal need for a long night's sleep. Showered and out of the motorhome by 5:30, I meet the sun sneaking over the flattest of horizons. Song sparrows chirp and flit in the bushes beside me. My 'walk around check' verifies we are ready for a full day's travel. Good tire pressures, full fuel and just half a quart of oil to add, radiator full, satellite radio and noise-canceling headset plugged in and water supply in place. With this sky, I certainly don't need my rain gear. As I roll on to the empty highway and shift into high gear, I find myself uttering a resounding "Yes! This is perfection." This is exactly what I imagined 10 years ago when I first started thinking about this trip.

Mt. Robson, B.C. Our home on the road for two and a half months.

But life is rarely that predictable. An hour or so later, I notice a thin band of clouds forming on the southeast horizon. My pilot voice says 'pay attention here.' Ever so slowly, but increasingly, the clouds build and begin to drift northward. I imagine our intercept in an hour or so at the same time that I realize my rain gear is miles behind me. I try calling Deb on the cell phone but there's no connection. The rain begins, lightly at first and then quickly ramps up to a full-blown thunderstorm. As I throttle down and pull over to the shoulder, an ominous clicking emanates from the bowels of the transmission housing. I'm in the middle of the prairies, chilled to my bones, and my tractor has sounded something uncomfortably like a death knell. Perfect days can be quite transient, so it seems. Three days elapse before my tractor is repaired and I'm back on the road again.

In addition to two major mechanical breakdowns, Ol' Red will suffer from other less grievous pains as well. Always, farmers and garage owners along the way are willing to re-arrange their schedules to effect immediate repairs. Kitchen doors swing open with offers of lunches and suppers. And throughout, farmers play a key role in helping with my photography. The limitations of shooting from my wheelchair become painfully obvious in a rural terrain. Deb and I quickly discover the means to transcend what seem insurmountable barriers — riding quads (ATV's) across rutted fields and over steep hills, and being raised up in the buckets of front-end loaders and forklifts for those semi-aerial views.

One element of my ride which completely surprises and angers me at times is the number — a small percentage I'm sure — of truckers who seem bent on convincing me Canadian highways were made exclusively for their benefit. Constantly checking my mirror, I pull over on the shoulder for overtaking traffic, yet numerous times I'm nearly rocked out of my seat as trucks roar by just a few feet away, needlessly blasting their intimidating air horns.

By the time I arrive at the Manitoba — Ontario border I have had enough. The shoulders of Highway 17 along the north shore of Lake Superior are noted for their general narrowness. Driving Ol' Red though this stretch means defying safety. That, and the increased weariness brought on by our over-ambitious schedule leads to a tough decision: I will ship the tractor by truck to Sault Ste. Marie, saving me four long days on a dangerous road.

As the miles slide by, it occurs to me, I really am a motorist of sorts. Just a very slow one. So I should be entitled to my creature comforts like anyone else. On a Sunday morning, at the outskirts of Windsor, N.S., I pull into a Tim Hortons drive-through. The young teen behind the take-out never so much as bats an eye. She stretches to her limit out of the window, over the rear tire, and serves up my morning brew. Maritimers have learned to take it all in stride.

On July 6 and approximately 6,800 kilometres later, I finally back Ol' Red into the salt waters of the Atlantic at the Dartmouth Yacht Club. Understandably, my feelings are mixed this morning. I am relieved the ride is over. I am physically spent and emotionally drained. Alexander McKenzie King once said that the problem of Canada is that we have too little history and too much geography. I have to agree. Those are long miles between our two oceans. But on the other hand, I regret saying good-bye to Ol' Red and our daily routine. Operate a machine long enough, share a few intense experiences together, and an impossible-to-explain connection begins to develop. It's far from a heart-warming human encounter but tangible in an odd sort of way. Today feels like closing another chapter of my life. I had dreamed of this venture for several years. And now, as I slide off the seat for the last time, the long ride is over.

Ultimately though, what dominates my memories are the farmers and their families we've met along the way. I treasure their stories, the stories that follow in this book.

I can think of no other profession that features more eternal optimists. One prairie farmer explained he had only one year in 30 when the season went pretty much as hoped — good weather, yields and prices. For 29 years, an unimagined depth of optimism carried him through much more trying times. One can't help but admire that kind of tenacity.

Farmers are perennial risk takers. Many farms today easily represent a capital investment of two to three million dollars. Yet the two biggest variables which affect their income — weather and market prices — are completely out of their control. Consequently, most farmers operate with slim net incomes relative to their investment. I saw a farm pickup bumper sticker in Manitoba which read: "I work to feed 300 people. My wife

With Ron and Cathy Bonnett and their cattle. Bruce Mines, Ontario.
"No," says the cow. "I've seen this before. Carl's going to fall.
Carl's going to fall. I simply can't watch."

works to feed me." Overall, more than 80 per cent of all Canadian farms have at least one family member generating income by working off the farm.

Are there any other professions where people work consistently harder than farmers? During spring planting and fall harvesting, it's not uncommon to put in 16-18 hour days for two or three weeks on end. Critical periods when inclement weather is forecast readily lead to 36-hour work binges. A predictable, 40-hour work week has never been part of a farmer's equation.

Most of today's farmers are first and foremost astute business people. The traditional farming part of their work — planting crops, harvesting, repairing machinery — has taken a back seat to skilled management. Today's successful farmer is marketing savvy, internet wise, and a smart decision maker. When my father farmed 50 years ago, provided the crops were planted and harvested within the required weather windows, there were few big decisions to make. Success today hinges not on who is the hardest worker but who is the best manager — primed with information and capable of making prudent decisions. Over half of all farmers entering the profession today have a college or university degree.

Ultimately, it is all of us, the consumers, who benefit from our farmers' commitment to providing us with high quality, low-cost food. Globally, Canada ranks third among nations spending the least amount of our disposable income (only 12.4 percent) on our food requirements.

Having been raised on a farm, I have always been aware of the effort and dedication required to begin the food cycle. I'm even more conscious of that now, having crossed this great land of ours and met farmers coast to coast. It's a privilege to introduce them to you through the pictures and stories in this book. I salute them. They truly are our unsung Canadian heroes.

If you ate today, thank a farmer.

Postscript

Carl and Deb photographing Kakabeka Falls, Ontario.
Today a photographer, tomorrow a 'has-been'.

The families in this book represent the diversity of farmers found across this great country of ours. Their names were chosen at random, from a variety of sources including personal contacts and referrals from agricultural writers and associations. Their operations vary across a huge range; from three acres to 12,000; from organic to agri-business; from family-run to corporate partnerships; from riding horses on the range to driving 530 hp, $380,000 tractors; from a newly created to a seventh-generation farm; from traditional farming to agri-tourism. The dramatic differences between the lives and work of just these families precludes any attempt to define, once and for all, "the" stereotypical Canadian farmer.

Photographing these families and their farming operations was a surprising challenge. First — and perhaps no surprise to the farmers — it seemed the weather conspired against us. Days of overcast skies and continuous rain made outdoor shots difficult. More than once we scrambled through ten-minute shooting windows between rolling thunderstorms.

Our pre-scheduled visits rarely coincided with the right moments to show the farms at their visual best. No sea of red cranberries floated in the Keefer farm bogs. No purple flax fields shimmered in a Saskatchewan sunrise. No combines inched across a 160-acre Alberta field of golden wheat. We were a full month into our journey before we found our first *perfect day*. The combination of a flawless summer sky and the hay harvest resulted in the photograph on page 150. How we wished for more days such as this!

There were also the small matters of time and energy to contend with. On a typical day, I needed to be on the road by sunrise, and then to face six or seven hours of driving. My partner Deb would travel ahead with the motorhome. Her farmer interviews were usually well in hand by the time I arrived sometime after lunch. Farmers were typically busy, trying to finish their daily chores as well as accommodate our visit. We were often limited to only two hours for interviewing and three hours for photography. In an ideal world, we would have planned for at least one full day with each family.

Within a week of our west coast departure, I realized our schedule was brutally over-ambitious. A pounding tractor ride followed up with rushed photo sessions, was already a full day. To that, add the daily regimen of downloading, editing and filing images, a couple of meals, and maintenance for both a motorhome and a somewhat temperamental antique tractor… What was I thinking? I can still recall the confidence with which I sat in my office, months earlier, map and calendar in hand, imagining a leisurely tour of Canada.

But we did it. We dug deep and somehow unearthed the resources — internal and external — to keep up with our schedule. Our arrival in St. John's, after two and a half months, found us exhausted. Yet we were also exhilarated, in a strange and quiet way. Over the course of years, a simple idea had grown into a "why not?" From there, to months of planning. Finally, the ride itself. And now, the book.

Was it worth it? Thousands of kilometres, a thousand cows, hundreds of smiling faces later? You bet!

Carl Hiebert

Donna and Ed Salle. "I can't imagine a better place for a morning coffee," smiles Donna.
The front door of the Salle house opens to a pristine and tranquil view of the valley and North Thompson River flowing in the distance.

Cowboys and Quads

BARRIERE, BRITISH COLUMBIA

A S YOU DRIVE UP the precarious dirt road leading to the Salle Ranch, you start to wonder if your imagination is playing tricks on you. You're surrounded by hilly, rock-strewn forests — the last landscape where you'd expect to find cows lollygagging between the trees. Mothers and calves lounge in the mid-morning sun, nuzzling each other. Bulls stand off to one side of their harem, appearing unusually mellow and content. Now and then, one looks over with its huge, compelling eyes that hint of a welcome smile. Fir trees, scrub brush, and B.C. sunshine are apparently an idyllic combination for *what might just be* the happiest cattle in Canada.

This is the high bush country on Ed and Donna Salle's cattle ranch where fast moving quads (all terrain vehicles) and cowboys are both part of the daily scene. For Ed, an afternoon of blasting off on his quad at speeds up to 60 kph is part of what makes his work exhilarating. He's at one with the machine as he easily tracks over rocks and steep inclines to move around the herd. What his quad lacks in grace, it makes up for in efficiency and low maintenance.

Modern ranchers like the Salles seem to have found a perfect balance. In the summer months, the herd is moved to crown land using a round-up combination of cowboys, horses,

"You can't believe how long it takes to rope," says ranch helper Derek Lenton. "I've been practicing now for 10 years and I'm finally nailing it more than half the time. If we could just get those guys to stand still, like the posts we used to practice on."

"Lady is 13 years-old and supposedly retired," says Ed. "But start up the quad (ATV) and that's her wake-up call. She's on the quad before you are, ready for work. Lady has always had an instinctive ability for handling cattle. She's the best boarder collie we've ever had. We could never do our work without the dogs."

and quads. Ed Salle is a second generation cattleman. "We are bush country ranchers with cattle and 1,200 acres. Our family had horses for 80 years but when it became more economical to use quads, we sold the horses. Margins are thin in our industry and the quads save us money and time. We do fence repair, maintenance and calving with the quads. When there's cattle work, we bring in the cowboys."

A man riding a quad is a huge contrast to the cowboys on their elegant mounts. Ed's hired hands wear traditional cowboy hats, boots, and shiny buckles. They ride with ease, steering cows effortlessly from one pasture to another. The cowboys are a reminder of the days when horses ruled the range. To see their horsemanship and wrangling skills is a true delight.

The cattle have different summer and winter pastures where they roam and graze freely. Spring is an exciting time of year: 200 calves are born into the herd and the Salle family goes on 24-hour on-call duty. Ed says it's his favourite time of year. "In spring, killdeer, robins, and cows are all here with their babies. It's a stunningly green, lush environment. The neatest

thing about our business is witnessing the natural progression of life. It is so interesting to see a cow make an instant transition. Before giving birth she's on her own, then some natural click happens and suddenly she is a serious-minded, concerned mom. No matter how many times you see that happen, you feel an appreciation for the cycle of life."

As the average age of a farmer in Canada is 49, retirement and succession planning is a challenge many families will likely face in the next 10 years. Ed's two sons, Tyler and Trevor, are now transitioning to take over the farm. Ed remembers when he took over from his dad. "I guess now that Trevor and I have partnered, history is repeating itself. I was the baby boy in a family of nine kids. I never wanted to ranch but when I was 24 years old, mom and dad were in their sixties, and we became partners.

"Those were different times. We rode horses to school, and did everything by hand. We canned our own vegetables and had a smoke house for ham and bacon. Mom was a great lady who loved the farm but worked hard. She had a wood stove

Farmers across the country speak of unpredictable weather patterns compared to years past. Summers are unseasonably hot, dry, or wet and winters generally much warmer. Agriculture, more than any other profession, is hugely impacted by climate. Ironically, Canada is thought to be one of few agricultural areas which will likely benefit from global warming, thanks to a longer and warmer growing season.

Sharp fangs and quick hoofs. Behind this display of seeming violence, both animals clearly understand their limitations. In a few seconds, it's over, the dust settles, and each contender walks back to its respective corner of the corral. The bark is still most often worse than the bite.

and used a wringer washer. It was funny — we also used that washer to wring out the peas from the pods," laughs Ed.

"My dad was kind of a quiet, community-minded man. He came over to Canada from Germany as a baby with his parents in 1912. It was quite a twist of fate: his family booked passage on the *S.S. Titanic*. But when they arrived in England there was a mix up with the luggage and they ended up on the ship *Corsican*. They weren't aware of the Titanic's sinking until their arrival in Halifax. Of course, friends and family back home had feared the worst and were frantic."

Donna and Ed have raised two sons on the ranch, and Donna has not lost any of her enthusiasm for ranch life. She says, "I love the cows, the space, and anything with speed! It was definitely a learning experience for me when I was a new mom with small children. At calving time we would take the kids out with us and sit them in the shavings. They'd sit there very contentedly, just watching us grown-ups work.

"I think it's the spouse's role to be the connector on the farm. I'm always the in-between person connecting the boys, their dad, and hired hands. I'm deathly allergic to horses so it's

the quads for me. I love driving fast and the free feeling of being on a four-wheeler or a motorbike. It is a great chance to forget all of your worries and get rejuvenated."

The average farm in Canada grew by 11 percent between 1996 and 2001. Ed says cattle ranches their size are definitely dwindling. "You've got to go big to survive. Right now cattle ranchers are struggling to make a living. If we could find a way to make it sustainable and a more consistent business, we'd all be better off. I hate to see: *Move over family farm, here comes the corporate farm.* When that happens, it's not about family anymore.

"It seems that rural people are aware of city issues, but urban people aren't familiar with our issues. It's in their best interest to educate themselves and become involved. People could make their supermarkets more accountable. When consumers buy Canadian beef, they are supporting their own community.

"I think people would be surprised at the depth of feeling a farmer like me attaches to my land and animals. A cow's life here is 10–12 years. I become connected with them. My cows know how to communicate with us — they'll tell us what they need. The herd is a part of my family."

The Passionate Farmer

VERMILION, ALBERTA

FLYING THROUGH A FOOT of oozing, stinking, coal-black mud is Ken Farkash's idea of a good day at work. He's heading out on his quad (ATV) to Grizzly Bear Coulee — a series of rolling hills straddling the riverbed flats. The prairie air is filled with the sweet musky smell of wild sage and the hills are a mix of tall grass, weeds, and occasional wild-flowers. You don't have to be a professional tracker to determine what kind of life roams here. Tufts of soft, brown fur float through the air and large droppings litter the ground.

As he approaches the valley basin Ken spots his grazing herd of 180 buffalo cows and their baby calves. Immediately sensing his presence, the herd tighten their stance and move closer together. Even in their shaggy, patchworked spring coats of shedding fur, these are majestic animals. At the moms' insistence, miniature sandy-brown sidekicks respond ever so slowly and saunter into the protected inner circle of adults.

Ken is a friendly, outspoken man who has nothing but positive things to say about farming. With great voice inflection,

Mavis and Ken Farkash "Taking the team out to visit the herd is my favourite way to visit the past. It's a slow ride, a chance to let the world go by, watch some wildlife, and time for Mavis and me to catch up together. I love farming but it's hard and stressful. This is the best thing I can do to stay balanced."

When the settlers first arrived on the prairies, upwards of 60 million bison pounded across these grasslands. Today, they number a fraction of that, found mostly on bison farms like Ken's. "I love them," says Ken, "They're a smart animal. This winter we had a huge amount of snow and there wasn't enough grass for them. They simply found a spot where the snow was packed hard over a fence line, walked over the fence to the road, and here was a two-kilometre line of bison heading back to the farm yard. They knew where the hay was. Now can you beat that."

dramatic body language and hand gestures, Ken speaks with a preacher's fervour. "On the farm there is always the opportunity to accomplish a dream. It's a great life! As a kid, I grew up in a really small farmhouse that was almost like a wooden granary. It wasn't insulated, and when the wind blew, the curtains moved!

"Most people who raise buffalo fall in love with them. You don't herd buffalo like you do cows — you attract them with oats. They are gentle, wild creatures. I've had buffaloes eat right out of my hand!

"We used to raise beef but switched to buffalo because they are a lot less labour intensive. I guess you could call them the lazy man's cow. You see, they're the kind of animal that is best left to do their own thing. They go out to pasture to calve on their own and need very little vet time, and no antibiotics.

"Buffalo are smart animals and what we call easy keepers. When a herd has eaten all they can in a pasture, they'll come to the gate to let me know they are ready to move on. We use four-wheelers to move them two to three miles between pastures. But it takes special care. I talk to them in a calm voice and try to entice them. And you know, I've never had an animal chase me."

In a sprint or marathon, Ken wouldn't be much of a match. Buffalo are three times as fast as a cow and his only hope to keep up would be on one of his high-powered quads.

The natural lifespan for a buffalo cow is about 40 years. They use visual and auditory signals and have a highly developed sense of smell. With a gestation period of nine months, they'll average about 28 calves in a lifetime. Bison are vegetation grazers and weigh in around 430 kg. Their brownish black fur actually contains a down which makes it surprisingly soft.

Historical accounts suggest there were 60 million bison in 1800. One hundred years later, there were fewer than 1,000 remaining. But with the reintroduction of bison to the plains by ranchers, farmers, and conservation agencies, that number increased to an estimated 500,000 in North America by 2005.

According to the Canadian Bison Association, "This reintroduction has been a positive development for the environment. Following the near total destruction of Canada's buffalo herds in the late 19th century, millions of hectares of the native grassland habitat of the prairies were sacrificed to grain growing. The prairie sod, with its rich mosaic of animals and plants was drastically altered to support the production of a handful of agricultural crops — primarily wheat. Today, bison ranching plays an important role in the preservation of the last remnants of native grassland habitat on the prairies by providing an economically viable alternative to cultivation."

The Farkash farm is located on 5,000 acres in Vermilion, Alberta. With the help of up to seven part-time employees to assist with the 300 animals, Ken also grows wheat and canola.

"I love the fact that there is always something changing, and we keep adapting. From one year to the next, I work hard to make sure that I don't repeat the same problems. And when need be, I find ways to compensate for last year's hardship. But no matter how bad things get, there's something to be learned.

"You know, your character isn't shaped by those rose-garden experiences. It's shaped by the challenges, and hard times. I think we farmers will keep finding answers to problems, and we'll continue to move forward.

"If I had one message to give to the public, it's that people need to realize we are producing high quality food at a low price — and that is a good thing for everybody! Agriculture has the best supported network in the world — we have technology, expertise, and government. We are not an island. I am so thankful to be able to call upon those services and support. I say that my bank manager is like a best friend. Sure, I pay a user fee, but it allows me to grow!"

Ken is a farmer who embraces life and clearly understands the importance of perspective. "We are able to capture the heat from a sun that is 150,000 kilometres away. There is such incredible order in our universe! Yes, life can be difficult but we have the option to put a positive spin on it."

*Golfers rate their courses, in large part, based on pristine views. Several of the Thornley's 13 U-pick
strawberry and raspberry fields back on to the Atlantic Ocean and should be similarly ranked as among the finest.
Picking a pint of berries here can take a long while as one is immobilized by the captivating shorescape.*

From Forest to Fields

IT'S A STORY that could perhaps only be told in Newfoundland: a young couple's dream of carving out a lifestyle *from rock and bog*. In 1979, city dwellers, Philip and Rhonda Thornley knew they wanted to live in harmony with the land. They wanted to integrate family and work, vision and passion. Looking back, 59 year-old Philip says it was a battle they were well equipped to fight.

At a time when many of their friends were leaving Newfoundland to pursue careers off the island, Philip and Rhonda decided that moving away was not an option. They were intent on making a contribution, and to them, leaving meant copping out. Instead, the couple took up homesteading on 60 scenic acres of forest land near Campbellton. With Philip's parents, Peter and Ruth, they combined their five university degrees and harnessed the family's knowledge of science, engineering and commerce. Their motivation was high and their vision clear — create a harmonious lifestyle, live with intent, and grow great berries!

The merger of Philip's geology experience with Peter's engineering background and Rhonda's passion for the environment, created a dynamic force. The family's first, self-built home was a geodesic dome with aspenite walls and a ladder for stairs to the loft. Philip and Rhonda had two small children, no electrical power and no phone. Rhonda washed diapers, made bread, and bathed kids in large bowls with hot water heated on a propane stove.

Looking after the farms 15 peacocks is just one of the many summer jobs for son Andrew. On occasion, U-pick visitors are more than just surprised to see exotic tropical birds in the heart of Newfoundland. One young woman, diligently picking blueberries, glanced up at the precise moment a curious peacock, only an arm's-length away, chose to stare straight into her eyes. The poor woman screamed, fled to her car and had to be convinced of the bird's harmlessness before she would return to her picking.

It was a daunting task. The Thornleys took on the challenge of a cold climate, land with poor drainage, and soil that required irrigation and frost protection. The farm was situated in a shallow valley with thin soils scattered on a bedrock base — in other words, a water-logged combination of sand and gravel covered with organic muck.

Their first initiative was a water-level-control project or as Philip says, "Managing the water that God gave us." This meant digging a network of ditches and ponds, and tackling the major job of creating a dyke system. It took years of physical labour using detailed aerial surveys, corrugated steel structures, and hundreds of loads of poured concrete — all painstakingly hand mixed.

What followed over the next few years was no less than miraculous. With limited funds and manpower, the Thornleys harnessed soil, sunlight, and water resources to transform their forests into fields. They built 15 kilometres of road, four geo-domes, and 20 acres of irrigation pipes to become Newfoundland's largest 'U-Pick Berry Operation'. Their Campbellton Berry Farm (aka Strawberry Heaven) now hosts 12,000 to 20,000 pickers annually.

Philip's decision to "live on the land" has resulted in one family's lifetime commitment and dedication. "We are first generation farmers. We built ponds and roads, picked rocks, bought every piece of equipment and built every house. I didn't want to fracture our lives; my vision was to create a marriage of work and family

One can only hope moose in central Newfoundland have learned to read, sparing themselves a 10,000 volt reminder (similar to touching a very large spark plug), that some desserts are simply off limits. A 560 kg moose can devastate a berry field in just one night.

tied together. We chose to raise our kids as little homo sapiens as part of the Earth, with space to use their imagination — and to join together their work, study, and play.

"We strived to do something meaningful with our lives," says Philip. "We built the geo-domes because of the efficiency — it's like a cat curling up to keep itself warm. And I like the idea of symmetry, taking one or two basic shapes and multiplying them."

Rhonda says that U-Pick is an economical way of harvesting fruit, especially since Newfoundlanders are born berry pickers. "I think most of our customers now are in their 50's or a little older. It's sad though, because we are losing our young people. Rural Newfoundland is turning into a senior's province." She also admits that having the public tromp around in her backyard on a regular basis took some adjusting to. "The first time

"We're never short of fresh berries for dinner around here," says Rhonda Thornley.
"It's about a five-minute walk from the field to my table."

we opened our fields, we felt invaded! It was scary. We had to lay down some rules: how to pick: where to pick: and how much you can eat for free in the fields. Now, most of our business is with repeat customers who understand 'picking etiquette'.

Describing their lifelong partnership, Rhonda says, "Philip tends to have his head in the clouds, and I tend to be the realist. He looks at the dream, and I look at the foot work. I guess we balance each other out."

Rhonda's love for nature started at an early age. "When I was little, I saw my dad cutting flowers, and absolutely freaked out. I thought he was hurting them. He thought I was crazy." While Philip is a purist who says the only way to eat a raspberry is warm, right off the bush, Rhonda disagrees. "No, no, no! Berries are good, any way, any time!" she laughs.

Of economic necessity, Philip has worked off the farm for 27 years. And like many farmers across the country the Thornleys have learned to make do with old equipment, unpredictable weather, and limited cash flow. They remain hopeful that *eventually* the tide will turn, and that a few good crop years will put them ahead of the game. Rhonda says, "Our farm has taught the kids to be environmentally conscious, but it's also definitely made them aware of the hardships of life. We could not have survived without Philip's income from jobs off of the farm.

"Looking back over the last six years ...Well, the last two were extremely poor. And before that, we had three mediocre years. But the one before that was a good one. One in six...," Rhonda smiles. "People ask what keeps us going and I tell them it's a combination of craziness and being too stubborn to give up! But I guess really, it's being in this environment, and watch-

Philip and Rhonda Thornley have lived their 32 married years well outside the traditional box. They designed and built their house and two storage sheds fashioned after the geodesic dome style popular in the sixties. "I love it," says Rhonda, "Except we could never tell our kids to go stand in the corner."

ing things grow. My idea of heaven is rows of lush grass pathways between acres of raspberry canes."

Philip offers a few words of wisdom he hopes his children have embraced from their experience of transforming forest to fields: "Take the risk of loving something! Our life here on the farm is the source of everything. And at the end of the day, I have the satisfaction of knowing: Man, that was a good strawberry I grew."

Loraine and Steve Lalonde, in a duel of "White Lightning" popcorn.

White Lightning and the Blues

STEVE AND LORAINE LALONDE are pioneers in the field of organic popcorn. As the producers of *White Lightning*, they are building a niche market on just three acres of popcorn. *White Lightning* has the three necessary ingredients that constitute great popcorn — great taste, a high popability rate, and it melts in your mouth. With such a winning combination, it is no wonder that the Lalondes are confident that they will sway popcorn lovers everywhere.

Growing organic is a new venture for Steve, who says the internet was his most helpful research tool in learning Popcorn 101. "I started growing popcorn because I wanted to do something totally different and have some fun. It has definitely been a steep learning curve. We're finding that growing a consistent crop is the key to success.

"We knew we had a great product when we bought commercial popcorn, you know—those *other* brands—and did our own taste test. There was no contest! Our popcorn won over all of our tasters. *White Lightning* has a creamy, nutty texture that sets it apart. We also grow *Tullochgorum Blues*, which is a unique, blue-kernelled corn that really catches people's eyes. It's blue when you put it in the pot, then it pops up white."

Popcorn is a type of maize (corn), scientifically known as *Zea mays*. Native American folklore told of spirits who lived inside each kernel of corn. These were quiet, happy spirits, who quickly turned angry when their houses were heated. The hotter their little homes, the more enraged they became until finally, they would shake the kernels so hard, they'd burst out of their homes and into the air as disgruntled puffs of hot steam.

Popcorn is actually the result of multiple, small-scale explosions that happen when the kernels are heated to about 205

You could argue that growing popcorn has the most potential of any crop. Popping this 25 kg bag of corn would produce enough product to fill about 10 bath tubs.

degrees C. Each kernel contains a droplet of moisture that is stored inside a patch of soft starch. When the moisture turns to steam, the kernels expand. When there's enough steam and pressure builds up, the outer shell pops, and presto! The kernel releases steam — and performs a contortionist act by turning itself inside out. Steve harvests his popcorn when the kernels are just at the ideal moisture content of 28 percent.

Steve's motivation to take on the organic challenge was partially to reduce input costs, but he also wanted to move to a more natural way of doing things. "Our transition to organic started in 1997 and the land was certified a few years later. It costs $1,000 a year for the Certified Organic stamp of approval. I never enjoyed working with chemicals. When I looked at what we were spending on inputs — fertilizers and chemicals — compared to revenues, I realized we were spending as much as we were making.

"I think being organic has allowed us to gain a better understanding of how soil works," says Steve. "And we are not so afraid of a failure now because we are receiving a premium dollar for what we grow. We have more of a buffer. It's funny, even some of the conventional farmers are looking at us now and realizing that it works. The organic industry is growing 20 percent a year and in terms of agriculture, there's huge potential out there."

"Along with popcorn, we grow soya beans, winter rye, grain corn and barley. We also run a poultry operation with 28,000 birds, and we use the excess chicken manure mixed with straw as compost here on the farm. It's part of the PAF, Planned Agro Fertilization, plan. We use one third of the manure we produce and we give the rest away to local farmers.

Farming organically, without the use of commercial fertilizers and pesticides, is more intensive and requires special machinery. This tine cultivator, a European design, makes its first pass through corn and soybean fields just a few days after the seeds have germinated but before they have emerged. As Steve carefully adjusts travel speeds and surface pressure, the tines gently disrupt the first crop of weeds but leave the crop mostly untouched.

up to $5,000, so we've learned other ways to grow and to minimize expenses.

"As farmers, we are up against a lot. I don't think city people really understand the business of what we do. They see the land holdings and lots of equipment, but they don't realize that the land is a tool of our trade. Land doesn't provide a direct income. Farmers are on the job 24/7, and if the weather turns against us, we are hit right in the pay cheque! We are impacted by things out of our control — like the rising dollar and the global market. Brazil, for example, has just recently come onto the chicken market. Five years from now, we will have a tough time competing. They have a number of advantages — a 12-month growing season, cheap labour, and a lower standard of living."

They'll use it as fertilizer on their fields. The nice thing about chickens is that the chores are a breeze. With our modern barns everything is automated — the work is a fraction of what it used to be.

"I learned my work ethic from my dad. He taught me how to go that extra mile. As a kid I worked along side of him and put in a lot of long hours. And when I grew up, he just expected me to stay on. He was the kind of guy who never took a holiday and just loved being on the farm. But I farm differently from the way he did. In his day, getting ahead meant expanding and buying more land. In our community, an acre can cost

The Lalonde's 115-acre farm is named *Tullochgorum*, Gaelic for *Green Hills*. Although the day-to-day operation is left to Steve, Loraine is a full partner and decision maker in the business. "You have to understand that when you marry a farmer, you marry the farm. I didn't grow up on a farm, but I was raised in a rural area. My first job was on my uncle's farm driving combines."

Loraine is an artist who specializes in murals and sign-making. She is the creative designer of the *Tullochgorum* popcorn brand's logos and product labels. She is also the first woman president of the Ormstown Agricultural Fair. The fair is home to her 2.5 by12 metre mural depicting the four seasons of agriculture.

"Steve and I have always worked together and have been our own bosses here on the farm," says Loraine. "I guess this is our little piece of heaven. It's the best place in the world." The Lalonde partnership is a blend of good business sense, creative initiative, and the ability to think outside of the box. And with that formula, it seems highly unlikely that the delicious secret of their perfect popcorn will stay a secret for long.

The Heart of a Farmer

WILLIAMSTOWN, ONTARIO

AGRICULTURE IN CANADA is big business. There are about 240,000 farms in Canada and approximately $34.2 billion annually is derived from the sales of crops and of livestock. But the business of farming is only part of the story. A farmer's occupation is also his lifestyle—where he lives, plays, and raises his children. Farmers are entrepreneurs who live and love within their business.

But when things go terribly wrong and tragedy strikes, how does the farmer separate his family life from his work, and his heart from the farm? In 2003, as Kerry McDonald lay in a hospital bed contemplating his own death, he struggled with questions that he had not explored before. If he passed, how would his family survive without him? Who would take care of his animals and the farm?

Kerry was the stereotypical dairy farmer who worked from sunup to sundown. As a sixth generation farmer, a father and a grandfather, his identity was defined by farming, and by his *need to work*. It was in this context that he now faced a life-and-death reality. "My heart was pumping only eight percent of the blood my body needed to live. It was almost hopeless to expect a heart transplant as my rare B+ blood type is shared by merely two percent of the population. The odds of someone deceased having this uncommon blood type and a viable heart were very slim," says Kerry.

Kerry needed a miracle to survive. Today, he is *living* proof that miracles can and do happen. Just after doctors told him he had only a few days to live, 'the call' came. A 39 year-old heart was available from a donor with the same blood type. "I grieved the donor's death for six months after the transplant," says Kerry. "Here I was, a 60-year-old farmer with the heart of a 38 year-old whom I'd never met. I am still very grateful for that gift and the technology that made it happen."

Eleven-month-old grandson Tavish. "We don't have a lot of toys like Nintendo games around here," explains Tavish's mom Tawnya. "Between the animals, the hayloft and the creek behind the house, the kids never seem to tire of all of the things they can do. They play outside whenever they can."

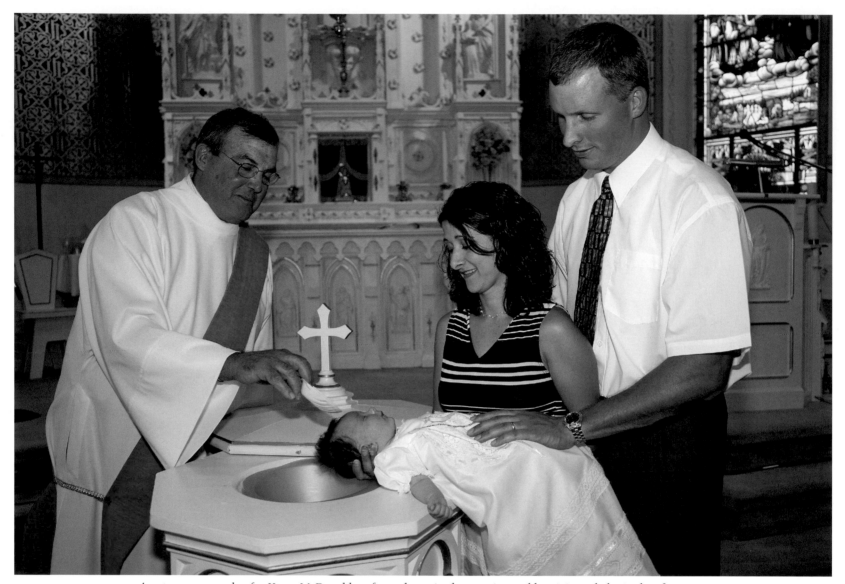

A not uncommon day for Kerry McDonald — farm chores in the morning and baptizing a baby in the afternoon.

Although Kerry appears to be the same man on the out-side, he will tell you that he is a different man on the inside. "When I was laying there with only a few days to live, I began to fight. I desperately wanted to live to see what my farm would be like in 10 years. I was so lucky. The transplant was a complete success and now, four years later, I am as strong as I was before. But since then a few things have changed. I don't spend every minute of the day working. I'm happiest just staying close to home with my family. I think having the transplant helped to bring me closer to God. I am a deacon at our church

and so far I've performed more than 50 weddings and over 100 baptisms. I especially enjoy the baptisms, and holding the little babies."

The McDonalds have 220 acres in Williamstown, Ontario. Their property is adjacent to the Raisin River, named after the wild grapes that grow in the region. They milk 62 Holsteins and grow hay, soy beans and corn for silage. For silage production, whole corn stalks are harvested while they are wet and green. During storage, the plant undergoes a fermentation process that makes it tastier and more digestible for cows.

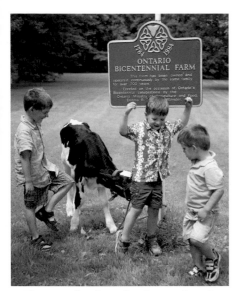

There are few farm families in Canada who can document their history for over 200 years. In an age of extended families living worlds apart, grandsons Lachlan, Ross, and Gavin will never question where they came from.

Kerry and Francine McDonald. "My great grandfather built this house about 150 years ago," says Kerry. "I was born here and could never imagine living anywhere else."

"My grandmother always said, 'Our farm has never been bought and it's never been sold'. Now, my son Tom and his wife, Tawnya, are responsible for keeping the farm books and taking care of the milking. Tom's a great farmer. As a youngster, he was out in the barn as soon as he could walk! He's very conscientious and he knows a lot about the technical side of farming — the feed, the mix, and the soil. What took me 15 years to learn, he got from agricultural college in just a few years.

"I think the future of dairy farming looks positive. As long as we continue to get a fair price for our milk and the supply management system stays in place, there will always be small family farms like ours. If I could say anything to people who aren't familiar with farming, I'd tell them that we farmers are just trying to make an honest living the best way we can."

Kerry's farm is a charming bicentennial homestead that has been in his family for 224 years and now accommodates the eighth generation of McDonald children. In a playground of tractors, hay bales, and dusty crops, they too are being imprinted with an awareness of the earth, the cycle of life, and the history of generations who have passed before them. Kerry's words of wisdom for these future farmers are short, but very sweet: "Live life simply."

Tom McDonald represents the seventh generation on the McDonald farm. "The future looks good and someday I'd like to see the farm passed down to one of my boys. That would make me happy."

Doug may be the only Canadian farmer to carry a steno pad while driving tractor — moments he guards as his 'think time'.
Always the innovator, he pioneered the world's first 'portable peach trees'. Winterized in a green house,
his 150 dwarf trees produce the largest, juiciest and tastiest peaches anywhere — and command a premium price.

Marshmallow Trees

WOLFVILLE, NOVA SCOTIA

DOUG HENNIGAR IS as forward thinking as they come. If his personal credo isn't 'Dare to be different', it should be. Among other innovate ideas, Doug is the mastermind of the 'portable peach orchard'. He has an orchard of 120, three-metre high trees, each living permanently in a little boxed environment. The trees spend spring and summer outside. Then, just before the frost hits, Doug moves them, one by one, to their winter greenhouse home. The result? The sweetest, most succulent peaches you've ever tasted!

And people are still talking about the time he made the six-o'clock news. Viewers were completely caught off guard when a CTV broadcast introduced Doug and his new hybrid of Mellow Trees. The TV footage showed rows and rows of an orchard, with hundreds of bare apple trees, covered with thousands of white fluffy marshmallows. The phone at the television station rang off the hook with callers who had failed to notice the date — *April 1st.*

Doug is a natural marketer and skilled businessman. When other farmers are turning away from the fruit industry, he's growing strong. For him, tractor time isn't about ploughing, it's about creative thinking. Every day when he's out in the field, he's armed with a notepad and pencil. As new ideas percolate, he expands his list of action items and creative ways to grow the business. He discovered a long time ago that to be successful, you have to be open to new ways of thinking, and be passionate about what you do. His advice to other business owners: "Don't think you have to be average. And don't be afraid to challenge the status quo!"

Doug and his wife, Heather, farm 80 acres of rolling, lush orchards of apples, peaches, pears, and plums in Wolfville, Nova Scotia. Wolfville is one of Canada's top regions in terms of ideal soils. The region also has the highest tides in the world and 10,000 acres of farmland protected by dykes.

This couple has a proven track record of taking on calculated risks and being one step ahead of competitors. They operate a U-pick flower operation, nature trail, petting zoo, and a farm market that sells their own produce and plants.

Cyril Whyte is one of Hennigar's seasonal fruit pickers from Newfoundland. "It seems city kids don't have the work ethic or interest in manual labour anymore," states Doug. "We're lucky folks from the island don't mind this work or we'd be importing seasonal labour from the Caribbean or Mexico, like most of Canada and the U.S."

The Hennigars, (left to right): daughter Kate, Doug and Heather. The sign tells only part of the story, for the Hennigars also operate a very successful agri-tourism business complete with petting zoo, flower garden, duck pond, tricycle track and picnic park. To say nothing of the 100,000 scoops of ice-cream dished out each summer at their road-side market garden.

"Agri-tourism is a big part of what we do and for us, that's about having fun with products and consumers. I think there is too much negativity and bad news in the media today, so we try to add humour to all of our advertising. We paint faces on our round hay bales, offer apple blossom tours and hay rides, or whatever it takes to make people smile."

Hennigar's farm produces 70 varieties of apples. "Our 'Honey Crisp' apples are probably the closest thing to a perfect apple that you can get. It is unique in looks — it's large, pinkish, and streaked with red and yellow. When you bite into it, it has a sweet, crunchy snap. Right now, we are producing 800 bushels of them a year."

But what about those wandering peach trees? "Our peaches are tree ripened, picked, and brought into the market for sale. We try to educate people about our fruit by sharing the genetic background and giving them nutritional information. We use one-half of the insecticides and chemicals that other farmers use by making use of beneficial insects. I think farmers have an obligation to protect the environment and we work at trying to understand that balance. We may be on the leading edge of that."

They are the largest hand-scooped ice cream operation in the province. "Last year alone we scooped 100,000 ice cream cones. We are one of the few retail outlets that allow consumers to eat *and* have a farm experience. About 400,000 people come through our place every year. Our motto is 'Eat closer to the farm.' We build on that difference by allowing people to come sample the fruit in our market. Then they have an opportunity to experience what we mean when we say, 'Eat local and you will taste the difference'.

Doug credits part of his success to belonging to a North American direct marketing group where he is surrounded by like-minded business owners. Participants openly swap ideas on how to become more effective and increase profit. "If you really want to be successful in marketing, you have to leave the farm. I've learned a lot by travelling to conferences and talking to other farmers. Members of our group are spread all over the continent, but we conference online, and it's instant, open

Fruit at the Hennigar farm store personifies farm-fresh, much of it picked that day. But consumer demands are fickle and Doug is always trying to stay ahead of demand. He presently grows over 70 varieties of apples on only 35 acres.

communication. My group has huge drive and hope — we are very optimistic."

"You have to be able to spot trends and react quickly because consumer appetites change," says Doug. "It's okay if people copy me, because I'm already onto the next thing, and ahead of the game. We've made a big difference in our operation by paying attention details. The newest, hottest apple on the market this year is Honey Crisp. Well, I planted those 12 years ago. Back then, everyone said, 'How can you take the risk?' My answer then, is the same as it is now: there's a bigger risk in not doing it."

Doug is a third-generation fruit farmer who says that rural Nova Scotia was the best place for a kid growing up. "Everyday was an adventure — it was a Peter Pan existence. My father was an extremely exciting man to be around. I remember being poor, but not in a negative way. It was okay because we had lots of fun just being kids.

"I studied wildlife biology at Acadia University but came back to the farm because of the opportunity. The transition of taking it over from my dad was super. His philosophy was, 'If there's an easier or better way to do something, then do it. Don't worry about me!' Since then, I've worked in the research side and in direct marketing, so I've had lots of freedom to experiment.

"The future of farming is difficult to predict. It's a challenging profession. The most rewarding part of my work is when I can bring an idea through to fruition and have it all come together. Passion is what drives something to another level, and I definitely have passion. I guess it helps to be a bit of a dreamer."

Karen and Ian Moilliet — and Willie. Sheep birth twins about 50 percent of the time. With triplets, as happens on occasion, there simply is not enough mother's milk. The resulting orphan lambs need intervention and special care. Willie fared best of all, being selected for training as a "lead sheep". Fitted with a leash, Willie quickly learned to follow a person, and the rest of the sheep simply followed Willie. Moving a flock of sheep couldn't be easier.

By Hook or by Crook
A Shepherdess protects her flock

VAVENBY, BRITISH COLUMBIA

THE SCENE IS IDYLLIC. A shepherd ambles along the meandering Aveley trails, surrounded by a flock of wooly sheep and accompanied by her two trusty guard dogs, Sebastian and Willow. The air is sweet and pure, and the land is just as pristine as the Moilliet family promises. Creeks and fresh-water springs sprinkle the meadows and woodlands. Tall, lush valerian and lupin flowers make it one of the most serene places found anywhere. This is the Aveley Ranch — a land of breath-taking beauty.

Tucked away in this secluded, little piece of heaven is a tiny shack where Kerry Moilliet will live for the next two months. Her barn-red, wooden shelter closely resembles an old fashioned wild-west wagon, complete with canvass roof. Her only heat and light will come from a rustic wood-burning stove and gas lantern. The indoor clothes line, hanging tin cups, and cast iron pots, all tell a story of a simpler way of living. This is where the resident shepherdess calls home.

"C'mon nanny, C'mon nanny," she calls, and waves her crook, herding her wayward sheep out of the bush. The flock must stay together, as predators like coyotes, wolves, cougars and bears are a real concern in this wild country. The setting is remote, but for Kerry there is contentment in isolation. "Someone has to stay and protect the sheep at all times. I enjoy working with the animals and simply being here. It's a peaceful place — I have lots of time to think and just…be."

The Aveley Ranch was established by Ian Moilliet's grandfather Tam, in 1906 in the North Thompson Valley. Spread over 1,800 acres with 1,300 ewes, it is currently the largest family-owned sheep operation in British Columbia.

Life at Aveley is a family affair for Ian, his wife Karen and their seven children, ranging in age from eight to 30. Ian is hopeful that farming will continue to be a way of life for his children, and for generations to come. "Growing up, I never considered doing anything else. I took the farm over from my dad, just as he did from his father. And I'd like to see my children go on making a life here."

"Living up here in the alpine meadows, our sheep have the best forage and freshest mountain streams. The only extra thing we feed them is salt," says Ian. "Maybe you can buy cheaper lamb imported from New Zealand, but you'll never match ours for quality. There's something to be said for buying fresh and buying Canadian."

The Moilliets are a tight family who pull together as a team. "We have about 900 births each year so come lambing season, there are babies everywhere," says Ian. "All of the children — April, Luke, Adam, Joseph, Vienna, Seth and Isaac — have a share in the work here. Vienna, for example, is only 14 but she is already the day lamber and looks for problems during birthing. My sister Valerie is the flock manager who oversees the overall well being of the flock, including alpine shepherding in the summer. The children are also responsible for doing house work and making a good number of the meals."

During lambing season when work days are extremely long, the family makes it a priority to spend time together. Each morning they sit down to porridge, bacon, eggs and toast. Breakfast is an important time — to pause for prayer, enjoy each others company, and talk about the day ahead.

Ian and Karen believe both work and play keep a family together. Karen laughs as she recalls meeting Ian at a bible study group. "The first time I saw Ian, he came walking up to the house wearing a cool western hat with a rifle slung over his shoulder. He seemed so mature, so manly. When I was young, I would tell people I was going to marry a farmer. It turned out to be true."

Karen says being at home on the farm offers parents many advantages. "During the last 20 years, I have home-schooled all of my seven children. It is a major commitment that requires a lot of work. And it certainly isn't for wimps, but the benefits are huge. It's a great way to teach your kids about the importance of relationships and character building. We hold school from September to March, and then they learn agriculture through being active on the farm."

Karen also dedicates a good part of her day to managing farm tours and renting out a charming log-cottage bed and breakfast aptly named Solitude. The private retreat is nestled in a scenic part of the property with a waterfall nearby. "The best part of living where we are is the solitude and of course, that's how the B&B got its name. We also offer a wide range of activities for visitors who want the opportunity to interact with the family and learn about lambing."

At only two months, Balto is small enough to have the sheep take care of him. Yet by the end of the summer, this Morama-Akbash guard dog will be the front line defense for the sheep wandering alpine meadows. Once grown, these dogs remain mostly solitary, sleeping near the flock. Their cuddly appearance belies a protective instinct charged enough to fend off coyotes, wolves, cougars, and even bears, several times their size. "Most animals get extremely annoyed with a barking dog," says Ian. "I've seen a little collie put a 70 kilogram cougar up a tree."

The farm's transition into agri-tourism evolved out of economic necessity. "Hard work sometimes just isn't enough," says Ian. "We are trying to figure out ways to spend less, but it's discouraging. What we do is simply not sustainable. Between rising costs and lamb prices that reflect 1966 prices, we are just getting by."

The Moilliets raise and breed Corriedale sheep for the dual purpose of wool and meat.

Each year, they sell 4,100 kg of wool to the Canadian Co-op of Wool Growers in Ontario for $1.54 a kilogram. In October, the lamb meat is sold to a central processor for $2.75 per kilogram.

Ian describes the ideal newborn as a strong-willed animal who will actively seek out its mother's udder. "A great little lamb is one that is in a hurry to walk and will stay with the flock. Eventually, she'll grow to be a strong ewe and produce 15 to 20 babies. Our lambs are born in April and then go out to fresh spring pasture. In the summer months, the sheep go to the high alpine meadows and will travel up to 35 kilometres. By the middle of September, they'll come back down from the mountain, and that's when we wrangle out the lambs and the whole marketing process begins."

Aveley is marketed as 'A ranch that is actively seeking consumers who have a healthy desire to eat naturally-grown foods. Lambs are finished (final feeding) in the heart of British

"When you've been up since six doing chores and lambing, you never turn down a chance for homemade scones — at any time of the day," says son Joseph. "As a family, we spend more time in the kitchen than anywhere else. During lambing season, we'll have up to 15 people around the table."

Columbia, where only mountain streams irrigate spring pastures and valley meadows. These animals have a healthy Omega 3 fat ratio. Scientific research has discovered Omega 3 fat to be highly present in grass-fed red meat.'

The last time you dined at your favourite restaurant and ordered lamb, was it Canadian grown? Ian suggests probably not. Often the lamb is brought in from New Zealand and Canadians aren't even aware of it. But the "buy local" issue is just one of the challenges this alpine sheep rancher faces. In 1966, an acre of land in Vavenby, British Columbia cost $200 and you could buy a kilogram of lamb for $1.10. Today, that same acre costs $1,500 *yet the average price of a kilogram of lamb is still only $1.10!*

The future of agriculture in Canada rests on the ability of families like the Moilliets to sustain a living both for their family and their flock. Like many farmers today facing an uncertain future, Ian and Karen ask only for what is fair — the opportunity to continue to work and to receive a fair price for their product.

Seventeen-month-old Mason's sandbox is pretty much as large as he — and his dad — decide they want it to be.
"He's such a happy child," sighs mom, "He'll play outside for hours."

Crop Scout

NORTH BEDEQUE, PRINCE EDWARD ISLAND

IT'S A TEDIOUS PROCESS. Squatting, standing up and walking two or three steps, then squatting again. But after 15 minutes in his dusty potato field, Andrew Waugh looks pleased. He has found only three tiny white worms that are Spud Island's number one enemy: *the European Corn Borer*. For now, his crop is safe.

Crop scouting is an arduous job. During June and July, Andrew spends up to six hours a day walking the acres of his family's potato farm. As part of the farm's integrated pest management program, scouting is the more labour intensive part of the process. It does not require any specialized tools, fancy equipment or technology. But it does demand keen eyesight, excellent observation skills and the ability to deal confidently with 'make or break' decisions.

Andrew spends a lot of time crouching in hot, dry mosquito-plagued fields. Combining the skills of a scientist and a detective, he carefully examines the stalk and underside of each leaf, looking for signs of disease or insect infestation.

This season, Andrew's main concern is one that other farmers on the island have already discovered. Andrew doesn't

The Waugh's new potato harvester is complex and unquestionably expensive. But it's also extremely efficient.
It readily digs and loads 500 acres of potatoes annually, the equivalent work of approximately 500 people digging with forks and shovels.

pull any punches when he describes the 'black, slimy moldy stink' of blight. "On the island this summer we're facing blight — a threat that could ruin our crops. It's a fungus that gets into new growth and deteriorates the leaves as it works its way into the stalk. If just *one* stalk gets into our storage bins … the loss could mean financial disaster.

"One of the main crop protectants we use is the chemical Dithane. It's applied by a four-metre-tall sprayer that, ironically, looks remarkably like a giant insect. Over the course of the next few months, the sprayer will be used in the fields every five to six days to prevent blight from penetrating plants."

Andrew learned scouting at 22-years old: he's also a self-taught mechanic and welder, responsible for maintaining much of the farm's equipment. He possesses a crop scout's biggest asset — the ability to problem solve. "When I was a teenager I spent a lot of time in the field watching and listening, and I learned as much as I could from the Ag consultants. The major challenge for a crop scout is to make the right decisions.

"If I find four out of 10 plants with egg masses, I need to take measures to save the crop. But it's a delicate balance: not wanting to miss something versus spending money on spraying.

On any given year, we can spend upwards of $100,000 on spraying."

Canada has a regulatory body that governs the use of chemicals on farms. Farmers like Andrew write a test every five years to ensure that they understand their proper handling. He says, "We need to know how to use sprays in various wind speeds, and how to take the proper measures to protect delicate streams and natural water systems. We'd rather not use chemicals, but at times it's necessary."

The winding driveway that leads into the Waugh farm shows just a hint of the rusty-coloured soil so distinctive to P.E.I. The scenic island, with its fields of red dirt and its gentle rolling landscape, has an excellent growing environment: clean water, clean air, and long cool winters that cleanse the soil. These are ideal conditions for growing potatoes, making P.E.I. one of the top producers in Canada's potato market.

Four families, along with Andrew's grandfather, homestead on the 500 acres that Andrew calls home. Andrew, his dad and two uncles produce an astounding 13 million russets, round whites and seed potatoes. Most of their harvest ends up as one of the nation's favourite fast foods — french fries.

As crop scout, Andrew is responsible for identifying two types of blights, ten plant diseases and eight pests.
These European Corn Borers can infest a field within a week and easily destroy half of the years crop if left undetected.

The Waughs, (left to right): Natacha, Mason, Andrew. A fresh-air, after dinner walk is as almost as close as stepping out the kitchen door.
No farm in PEI is more than a few kilometres from the ocean, making it by far the most picturesque farming province.

Andrew has been helping out on the farm since early child-hood — he wrote his tractor license at age 14. With his wife Natacha, and their baby son Mason, Andrew has taken over the original farm house he grew up in. He says that farm life can be difficult on young families. Of the day-to-day uncertainty of farming, he says, simply, "It's been hard on us.

"We wish Natacha could stay at home to take care of Mason, but we need the income she earns from working off of the farm. It seems that consumers are easily persuaded about food choices. For example, a fad diet a few years ago recom-mended not eating potatoes. Our potato market took a big drop for about 18 months. That kind of thing puts the farmer in a vulnerable position.

"If in 10 years our family is still farming I will be very happy. But that worry is always in the back of my mind. As a young farmer, I've thought about my options. Obviously I'm not in this to get rich. I know that I could make more money doing something else, but then I wouldn't be doing what I love to do. There are more important things in life than money. Just as long as I can continue to farm and put food on my family's table, I'll be happy."

Allen's likes are simple — hockey and pigs. "Pigs are very curious animals. When I sit in the pen, they love to check me out because I'm a stranger to them. They like nibbling on my boots and clothes and if I'm not careful, sometimes they'll catch my skin. And that hurts!"

Sell the Camaro, Cut Your Hair, and Take Care of these Hogs!

WHITEWOOD, SASKATCHEWAN

IN 1960, IF YOU LIVED on a typical family farm, you might have had 500 chickens, eight milking cows, several fields of grain and hay, a family garden and a few pigs. You probably had four hungry mouths to feed, so raising sows for their litter was paramount for survival. The price of pork was $1.20 a kilogram. More importantly, the farmers share of the consumer pork dollar was 62 percent.

Today, the average hog farm has 500 sows, and the price of pork is $7.50 a kilogram.

And the return of the consumer's dollar to the farmer? Only 30 percent. As a consequence, farmers have been forced to go big. A visit to the Kilback Stock Farm (KSF), a modern pork production facility, is an eye-opener.

KSF, owned by husband and wife team Allen and Denise Kilback, specializes in processing semen, breeding sows, birthing piglets, and selling the breeding stock in Canada and United States. Overseeing an astounding 40,000 hogs annually is an enormous challenge, but it's one that Allen Kilback was groomed for. "My dad bought five quarters (800 acres) of land and gave each of his sons one quarter. When I turned 18, he bought 14 pigs and said to me, 'Allen, sell the Camaro, cut your hair, and take care of these hogs. YOU are responsible for them!' At first, I hated working with pigs. I guess I had to go through a learning curve. But I quickly learned that pigs are very intelligent. I can now say that I'd

It might be argued that bio-security is tighter on farms than in our health care system. Hospitals sent patients from one hospital to another during the SARS crises. A hog farmer would never think of shipping a pig from one barn to another during a disease outbreak. Showering and a complete change of clothing are required each time one enters and leaves this barn.

rather work with pigs then people … and, that says a lot because I'm a real people person!"

As you approach the barn, the first thing you're struck by is a thick, pungent odor. There's nothing subtle about this smell: it clings to the back of your throat. The 'No Entry' signs indicate the enforcement of strictly regulated access. Bio-security regulations require all visitors to shower and change clothes both entering and exiting the barns.

After showering, you enter a long narrow hallway which leads to numerous, smaller, barn-like rooms. Each of these units houses pigs in the various stages of their growth cycle — commonly referred to as a 'farrow to finish' operation. The pens are clean, have controlled lighting and are kept at a constant humidity and warm temperature.

The farrowing barn houses the pregnant sows who give birth to 85 babies every day. The average litter size is 11, and a healthy piglet will weigh about 1.36 kilograms. Soon after birth, mom grunts to her litter, rolls over, and then allows the litter to feed. Each piglet has its own dedicated teat that it will return to each time it nurses.

Denise Kilback spent years as a midwife in the farrowing barn, helping over 100,000 piglets come into the world. "Just after Allen and I were married, I worked in the farrow and helped the sows and babies. As a new mom myself, I could relate to the nurturing side. I took our newborn, Ashley, into the barn and put her little basket in a

The Kilbacks, (left to right):
Allen Jr., Allen Sr., Ashley,
Arielle and Denise

safe spot in the corner so I could help the moms during long labours. Sometimes, you have to remove a film and mucus from the baby's mouth and nose. I had one little runt I called Peanuts. He weighed only one and half pounds and fit inside of a coffee cup."

The newborns stay with their moms for 21 days before being shipped off to the nursery. The nursery is sectioned into smaller pens filled with litters of scurrying pink weanlings with soft coats and straight ears. Although some piglets have straight tails while others are curly, they all have four toes on each hoof. They are timid but curious creatures who will happily nibble anyone who enters their pen. They'll stay there for five weeks before being moved to the grower-finisher barn.

The pigs remain in the finisher barn for another 16 weeks until they reach market weight, about 116 kg. Almost every part of the pig is utilized. Bones and skin are used to make glue, gloves, footballs, and garments. Hair is used for upholstery and

artist's brushes. Perhaps surprisingly, pigs are a source of nearly 40 drugs and pharmaceuticals such as insulin. Pig heart valves are used as implants for humans.

KSF promises its breeding stock will increase size of litters and birth weights, improve feed efficiency and reduce the number of days to market weight. When 'gilts' (unbred female pigs) or sows enter their reproductive cycle, they are bred using only artificial insemination methods. A sow's pregnancy is 115 days long and she will usually have two litters of piglets each year. Canada annually produces about 31 million market hogs and exports pork to approximately 100 countries.

KSF's breeding stock is pre-sold and the slaughter meat is grown on contract for Maple Leaf Foods. Allen defines his role at KSF as a businessman who loves marketing. "I have a tremendous passion for agriculture, but I am trying to move us away from agriculture to a sustainable, profitable lifestyle. We have developed a niche market with our breeding stock and are

Denise Kilback never imagined one day becoming a mid-wife — to approximately 100,000 new births! "As a mom myself it seemed natural for me to nurture newborn pigs. In just two hours or so, you have a new litter of 11-12 piglets. They have hardly any hair and are very smooth and cuddly. You dry them off and get them under a heat lamp to make them as comfortable as possible."

heavily invested in technology. We use 'Pigchamp', a computer program that monitors factors influencing breeding perform-ance. Tied in with the science of PIC (Pig Improvement Canada), we're on the cutting-edge. PIC is one of the largest genetic companies in the world: we pay a royalty to use their genetics and have a complete DNA history on our animals."

Allen's long, busy days are no longer spent on a tractor or in the barn. He hopes his 80-hour weeks will become more controllable when he brings on another full-time manager. "I start planning the day at six, and by seven I'm on the phone conferencing with my three barn managers. Two days a week, I'm working in the office, but the balance is spent visiting other farms that are using my stock. This is a relationship business."

Allen sees a future of growth, and challenge, for the hog industry. "The biggest challenge we face is that the hog market is cyclical. It's feast or famine. We try to minimize risk, but the Canadian dollar is affecting our business tremendously.

"In the beginning we did everything ourselves, but now we are trying to create an organization that runs itself. I think we hog farmers will continue to have to produce for less because there will be more product on the global market. Fortunately, both China and India are consuming more pork and that is where the growth will be. Consumers should be telling farmers what they want," says Allen. "If they want more grain-fed ani-mals raised in a humane way, they must be vocal about it to supermarkets and restaurants"

Success in farming, as in all other areas of business, comes from executing a long-term business plan and, as Allen discov-ered for himself, understanding the power of marketing. "I think farmers need to look in the mirror and ask themselves, 'How do I fit in this food chain?' Farming is a business full of risk and marketing is everything. We sell 700 pigs weekly, but it took us 10 years to get here. Now, we're sold out and booked one year in advance."

*The Culbersons, (left to right): Leanne, Tammy, Daniel, Ross, Kristine and Rachel. An obvious advantage of rural living —
the potential for all the elements of a child's ideal playground…a farm pond, tree forts, skating rink, and family campfires.*

The Corporate Potato

WOODSTOCK, NEW BRUNSWICK

WHAT IS THE PERFECT POTATO CHIP? Crispy? Thin and salty? Or thickly ruffled with just a hint of oil? Potato chips were discovered quite by accident. The credit goes to chef George Crum, working at a resort near Saratoga Springs, New York in 1853. One of Crum's patrons, or so the story goes, sent his French fries back to the kitchen complaining they were too thick and too soggy. Crum retaliated. He cut up his fries so thin and fried them so crispy they couldn't be skewered with a fork. But the plan backfired. The fussy customer loved the tasty, paper thin potatoes. Soon after, *Crum's Saratoga Chips* were a regular item on the menu.

Growing potatoes for potato chips is much more complicated than one might expect. According to the 'McCain' contract that New Brunswick farmer, Ross Culberson maintains, *the perfect potato is defined by size, and its ability to consume a precise amount of oil in a limited cooking time.*

"Four pounds of our potatoes will make one pound of potato chips," says Ross. "We are on annual contracts for a total of 50,000 barrels of potatoes. In August, we'll dig about 300,000 pounds of our Atlantics to ship to the Hartland plant. The very next day, my potatoes will be cooked into crispy chips and bagged."

Ross and son Daniel, age eight. A generation ago, most Canadian farm families had their own garden and packed away a good part of their winter food supplies in freezers and cold cellars. A more wholesome food chain, perhaps, but one that has been largely replaced by convenient weekly drives to the nearest supermarket. An older generation still reminisces about smoked hams, jars of preserves — hundreds of them — and potatoes fresh out of the field.

Daughters Rachel and Leanne will never question the quality of their potato chip snacks. Their parents' farm supplies the McCains plant just a half hour away. Thanks to a detailed tracing program, from food source to bar codes on the retail chip bags, the girls can actually verify that the chips they are eating come from their very own farm.

The Culberson family story is one of three generations, passing on a business, a lifestyle, and a contract. Ross inherited the potato contract from his father, who inherited the contract from his dad. "My family has a partnership with McCain. They help farmers like me keep up to date with new technologies, processes, and concerns like food safety. Traceability is a huge issue now. At McCains, every bag of potato chips is coded with a series of numbers and at any given time, they can trace when and where that particular bag came from. I do my part by keeping traceability records with dates and types of chemicals sprayed and faxing this information to the corporate office."

Ross says his family are loyal potato lovers who happily eat them at least five times a week. And as tubers (root vegetables) are nutritious, fat free and high in fibre, they're an excellent choice for an inexpensive and healthy meal. "The Russets make good French fries. And the Snowdens will store for a long time and make a nice white chip. But my favourite is the Atlantic.

They are the best table potato. When boiled, they have a nice moist, creamy texture, and the peeling just washes right away!"

The history of the potato dates back hundreds of years and is rich in folklore, politics, art, and even fashion. It is said that Marie Antoinette made potatoes a fashion statement when she marched through the French countryside wearing potato blossoms in her hair. Vincent Van Gogh painted a number of still-life canvases devoted to the potato. Writing of his famous *The Potato Eaters*, Van Gogh expressed his respect for the peasants he depicted: "I have tried to emphasize that those people, eating their potatoes in the lamp-light, have dug the earth with those very hands they put in the dish, and so it speaks of manual labour, and how they have honestly earned their food."

Ross and his wife, Tammy, live just minutes from their 450-acre farm in the rural community of Woodstock. "My family now lives off of the farm so our personal life is kept separate. Things were different when I was a kid. I remember being six years old, out in the field picking rocks, when my uncle sat me on the tractor. He didn't tell me how to drive, he just pointed me in a direction, and I went for what seemed like forever. Those were the days when everything was done by hand. Now, our work is more complicated, tractors are bigger and the farm yard can be a dangerous place for children."

Tammy takes care of the farm's financial books, works full-time, and is mom to four children. "In the busy seasons, Ross is gone from sunup to sundown. That's when I become a single parent for a while. He has a good business mind for farming, but when he takes on something, he gets tunnel vision. Farming can be hard and when things get tough and don't go well, he becomes quiet and takes it inward. The kids and I are here to support him.

"Ross is faithful to the McCain Company and is adamant about us buying Humpty Dumpty chips. But you know, if I did buy another brand, I'd make sure that I got rid of the evidence!"

Ross describes the perfect potato plant as dark green with lots of blossoms. "The Atlantics are seeded about nine inches apart and each plant will produce six or seven potatoes. We use

"Kristine is only two but she's already a willing helper," says her proud mom.
"She particularly likes water and when I'm doing potatoes, she insists on standing on a chair by the sink to help with the washing."

a top killer spray which causes the spud to form a skin. This forces the top foliage to die and speeds up the drying process. The rotations of my crops are usually spuds one year, followed with barley, with rye-grass underseed to provide good organic matter the next year. And then every second year I grow potatoes in that field again."

Ross' work day starts at 6:30 when he checks the rain gauge and potato storages to ensure that fans and humidifiers are working properly. By eight, he's in the field. "During harvest, my days go to about nine o'clock. For nine hours, I'm on a har-

vester that's loud and dirty. The ground is rocky but the spuds always come up. Because our weather window here on the east coast is tight, we try to harvest in 20 days. Lately, it's been raining a lot. Last week, she was some muddy picking up those potatoes!

"I didn't always want to farm; I thought I wanted to be a nine-to-fiver. But I came back because I wanted to be outside and work the land. It's quite a thing to be able to go out and stick a seed in the ground and then watch it grow. You're making something … not just putting in time."

The Splendour of Soy

A COMFORTING AROMA OF cinnamon, rosemary, and nutmeg wafts through the air of Elaine Edel's Manitoba farmhouse. Resting on the counter are cooling racks filled with the day's baking: pies, waffles, muffins, and the most divine-looking breads. Her kitchen resembles a lab with its grinders, dehydrators, mixers and other tools of the trade. Elaine is not only a chef extraordinaire; she is a pioneer in food science and author of seven cookbooks.

Necessity really must be the mother of invention, at least according to Elaine. Years ago, when she was a young mom to five boys, her family found themselves overwhelmed with financial burdens. The ingenious stay-at-home mom rose to the occasion. Elaine contributed to the family income by using her culinary know-how and working magic in the kitchen.

Elaine's story begins when she was a teenager. "I met my husband, Melvin, at church when we were both 16. I immediately put my eye on him and I thought, 'He's not getting away!' We married soon after, began farming and started our family. I stayed at home with the boys, and spent a lot of time in the garden and cooking. I loved to cook and was always trying new things in the kitchen.

"In 1988, we had one of our worst crop years ever. We had a drought so severe it only produced two bushels of wheat to the acre. It left us financially stressed to the max. We had five kids to feed and needed to find a way to survive. That's when I knew I had to do something and cooking was what I knew best. I began catering from my home. I was always creating new recipes so I thought I'd put a few together into a cookbook."

Since writing out those first few recipes, Elaine has spent more than 15 years sharing her passion for cooking and her knowledge about food. She has worked with the University of Manitoba exploring the uses of soy. She's made over 10,000 loaves of whole grain breads with soy, flax, barley, rye and wheat. Elaine no longer cooks for survival nor just for pleasure — now she teaches consumers how to bring grains and plants from the field into their kitchen.

Under the pen name *Winnie*, Elaine writes recipes and gives tips for both the novice and the experienced cook. There are few experts who could match her knowledge of soy and fewer chefs who could claim to have created as many original recipes from soymilk, soycream, soynuts and soyflour.

Her book *Soy Satisfied*, offers readers 200 soy-based recipes and explains the health benefits from foods rich in this alternative protein source. "This little beige seed is so plain and so tiny, but yet so very powerful and nutritious. Soy improves the texture and flavour of foods. It provides high quality protein that is full of vitamins, minerals and fiber with no cholesterol or saturated fats. It is one of nature's most efficient foods. Soy provides a high level of nutrition at a relatively low calorie cost. Of all the seeds that have appeared on our table, soy is definitely the greatest!"

Elaine has not limited herself to recipes with soy. In her book, *Cooking without*

With a particularly wet spring, Melvin inspects his soybeans for insects. "Farming is tough," sighs Melvin. "We have to work harder and longer hours these days just to break even. How are my sons ever going to make it?"

Elaine and Melvin Edel

Mom, she focuses on young people fending for themselves for the first time. The recipes cover the basics from pasta to Taco Burger, but also provide invaluable household hints. Did you know, for example that "Egg white will remove gum from almost anything?"

Elaine and Melvin have been married for 44 years. Such longevity provides a level of comfort and loyalty that can only come from being lifetime companions. As Melvin says, "Elaine and I keep each other going. We're perfect together. We make it work."

The Edels farm with two of their boys on 3,200 acres in Morris where they grow soybeans, canola, oats, sunflowers and wheat. Asked to describe the region's farmland, Melvin says, "It's flat!"

You don't have to scratch the surface too deep to know that Melvin is a hard working farmer who misses the old days, and *the old ways*. "Back in the mid 70's you didn't have to work so hard to get ahead. We carry more debt now and the prices we get for our crops are way too low. We have to work longer and harder just to break even!"

Cats are to farms what jam is to peanut butter.
When granddaughter Gena comes to visit, she first checks the shed to say hi to the cats before greeting grandma and grandpa.

Hundreds of recipes and seven cookbooks later, Elaine is still experimenting in her kitchen — sometimes with the help of her three-year-old granddaughter Karly. "Right now I'm working with hemp seeds and wheat sprouts. But who knows, tomorrow it could be the healthy pomegranate. I can't imagine ever running out of ideas."

The Edels are frugal people who live wisely. They drive older vehicles and rarely eat out or take holidays. As Melvin says, "We're not fussy about things like that." He has farmed for close to 50 years and remembers when life was much less complicated. "My boys just bought an air seeder with GPS and I don't even know how to use it. Today, farmers are inside working on their computers, while I'm still out in the barn feeding my cows.

"I start my chores around five in the morning by taking care of the cattle. Then I head out to bale hay. I'm an older farmer and maybe I have my back up a little about the way things have changed. But you know, if we don't get a decent crop this year, my boys will be working out on the rigs in Alberta."

Melvin is a third-generation farmer, who at 64 years old is going strong. According to Elaine, he still does more than his share of field work and combining. He has endured much in his life and definitely earned his crustiness! "My brothers took me out ploughing when I was just five years old. Over the years, I've been kicked by cows, hit on the head by a fallen tree, and hit by a grain auger. I have broken a leg, ribs, and had more bruises than I could count. I was even hit by lightening once. I woke up hours later and spent four days in the hospital. And ever since, I haven't been able to stand hot days. But I still love the way lightening lights up the sky!

"What do I think about during all those hours when I'm on the tractor? Well, when I'm not singing or whistling … I think … I wish I could win the lottery. And then I think about the way it used to be and farming in the old days."

Organic Ann

ST MARYS, ONTARIO

STROLLING THROUGH 'Organic Ann's' vegetable garden is like stepping into a gigantic salad of bitter greens. A sharp aroma of radishes and onions permeates the air. Filtered sunlight dances through the field of oversized spinach, baby bok choy, and jewel-like emerald lettuces.

Gardeners from novice to pro, and a loyal following of discriminating buyers seek out Ann's knowledge and her produce. The attraction is obvious. She has the giggle of a young girl, the wisdom of a master gardener, and she's absolutely passionate about growing vegetables — *organically.*

Ann Slater was raised on a farm in southern Ontario by parents who instilled a sense of environmentalism. "I took over the family garden when I was only 15, and even then my produce was organic, just not certified. I don't know how to grow with chemicals — I never learned. My dad stopped using pesticides on the farm in the mid 1970's. When I realized there was a demand for great-tasting fresh vegetables, I set up a card table on a main street in town, and sold whatever I had grown that week."

After high school, Ann left home to study biology at Trent University because, as she says, "I was already a farmer and wanted to learn something else." But just before receiving her degree, she was reminded of her true calling. "I was near the end of my studies and had an opportunity to apply for *the* job. It was a great position as a biologist. But my only thought was, I can't, I have to go home and plant! And when I couldn't get that thought out of my head, it became clear to me that farming was what I needed to do."

Ann has achieved the epitome of intensive farming with her small market garden. As soon as one crop is harvested, she plants another. Hardy vegetables such as beets or Chinese cabbage, are frost resistant and are harvested as late as December. In typical cash crop farming, an acre of corn might net a farmer $70 an acre. Ann will generate at least 150 times that amount per acre!

After university, Ann returned home to the family's dairy farm. She expanded her market garden to one and a half acres of land, and relocated her card table to a more permanent home at the St. Mary's Farmers Market. She now supplies vegetables to two local restaurants and operates a weekly vegetable basket delivery program. She says her clientele has grown mostly through market traffic and by word of mouth. "As the market grew, I grew. Each year I saw the demand for organic produce grow. Now, I'm barely able to keep up! My customers are people who have environmental and health concerns. They buy from me because I'm certified organic, I'm local, and they know my produce is fresh."

Ann says that as a young girl in the early 90's, her aspiration to farm

was not met with enthusiasm. Public perception of the average farmer at that time just didn't include an energetic, forward-thinking young female. "My biggest challenge wasn't being the 'rebel organic market gardener'; it was gaining acceptance as a female in farming," she says. "People would tell me, You can't be a farmer. You have to marry a farmer.

"During the growing season, I work from sunrise to sunset — from 12 to 15 hours a day.

It's hard work, but I love it. Fall is definitely my favourite time of year: that's when those 30-degree, humid, sweaty days have passed, and my tomatoes are at their best. I *love* eating tomatoes! And then at Christmas time when everyone here is still working, I'm lounging in bed and snoozing," she laughs.

"People often assume that you will yield less by growing organic, but in my experience, conventional vegetable farmers do not get near the production I get. At peak time, my field is covered with so much green, it's difficult to even walk through the garden. I harvest two or three plantings a year and grow 30 types of vegetables. It's pretty intensive farming. As soon as one crop is off, I immediately plant another.

Ann and her brother Stew make a unique farming duo. Not only do they work on the same farm, but both operate organically — albeit in different products. Stew runs a 150-acre dairy farm with 25 cows. Ann rents only a small corner of it — one and half acres — for her ecological market garden.

"I have no shortage of fertilizer," she smiles, "That comes from the manure producers — my 10 sheep. There are a number of ways gardeners can work in harmony with nature, like working white clover back into the soil as an effective green manure. There are also many ways to attract beneficial insects — you can let a cilantro crop go to seed to attract lady bugs, lacewing butterflies and paradise wasps. To combat the bad insects, I cover the plants with floating row covers. These are thin, polyester cloths that you anchor down on the edges. Rain and sunlight can penetrate, but beetles and such are kept out.

"It surprises people when they find out that my planting season never ends. I use a 'hoop house', a small greenhouse structure that extends my growing season. Unlike most farmers, my harvest season starts in April with the spinach and leeks, and just keeps on going until late fall."

Ann lives in the loft of the family's original homestead, a century home built by her great-great-grandparents. Ann shares the house with younger brother Stew and his family; Ann's parents, who are retired, live in a granny flat downstairs. Stew took over the family organic dairy operation a number of years ago and is one of Ann's great supporters. "She is knowledgeable and well-connected in the organic industry. The fact that she produces something that our family can use on a daily basis is really helpful. When you taste supermarket carrots and then Ann's carrots, it doesn't matter what the price is. You're going to buy Ann's because there's absolutely no comparison in how they taste!"

Stew says the organic dairy industry is growing because the nutritional benefits that come from feeding cows a diet of hay and pasture is something that many consumers are willing to pay for. "The term organic, when it applies to dairy, means a lot of things. The most important thing is that our animals have a more natural diet than what's found on most dairy farms. Our animals are outside all summer on pasture, and although they are brought into the barn during winter, they're exercised daily."

During their teen years, Stew and Ann belonged to 4-H Clubs. This Canada-wide organization offers young people hands-on programs to learn about their environment, in accordance with their motto *Learn to do by doing*. "I went to 4-H from age 12 to 18 and always had a project on the go," says Ann. "We had our own goats, chickens, and pigs to look after. It was a great way to learn about animals and nature, and to gain leadership skills."

Ann's ability to lead is evident in her role as president of the

"*I grow about 30 types of vegetables and can barely keep up to demand," states Ann. "My customers are people with environmental and health concerns. Pull one of my carrots out of the ground and you will not find anything fresher or tastier.*"

Ecological Society of Ontario. Along with creating public awareness and a community of like-minded individuals, the organization's mandate is to promote ecologically sound agriculture. As president, Ann has been very vocal in speaking out on behalf of consumers. "The people I meet at farmers' markets who buy organic are well-informed. They're concerned about the widespread use of pesticides and the risk of contamination of our water, air, soil and food. Pesticide manufacturers seem to be worried about the potential loss of sales opportunities as consumers choose to buy more and more organic food, since organic farmers do not use their products.

"If people want to protect our environment, they should consider buying organic food. They can be assured that organic farmers are not adding more potentially harmful synthetic chemicals like pesticides and synthetic fertilizers to our environment."

Ann's motivation to farm organically comes from listening to an inner voice and following her passion. Her need to work 'soil to seed' and nurture plants to become their very best, is a gift. "I love interacting with nature and I can enjoy the unexpectedness of it all because I am not around anything that might be harmful," Ann says. "I can be out in the garden with my little niece and pull something out of the soil for her to eat right there. It's completely safe.

"I have an image in my head that I saw when I was a little girl. It was of a clear-cut area in Malawi, Africa. Even though I was only a child, I knew that what I was looking at wasn't right. To me, it's wrong to put deadly things into the earth. As individual people, we can only look after one small space. But if we all take care of that one small space, we will make a difference on the planet."

Canada's Oldest Farmer

BRIDGETOWN, NOVA SCOTIA

A S THE EARLY MORNING SUN begins to brighten the Annapolis sky, Robert Chipman waits patiently in the yard for his dad. As predicted, 93-year-old Eugene slowly rolls up in his purple minivan exactly at 7 a.m. He's here for a quick chat before he and Robert start work. The feisty patriarch still retains 51 percent of the family's 300-acre mixed farm. Eugene just might have the honourable distinction of being Canada's oldest working farmer.

Eugene grew up on a farm in a small, rural community, but unlike most of his friends in Tupperville N.S., he chose not to follow in his father's footsteps. "After high school I went to McGill University to get a science degree. I then immediately left for Guatemala to work in tropical crop research. A few years later when my dad suffered a heart attack, I had to return home but I continued working as research scientist in Kentville. I drove home weekends to take care of the family farm.

At 93, Robert's father Eugene may well qualify as the oldest farmer in Canada. "During harvesting, I'm usually up by six, and in the field by seven, looking after the pickers. I've had my hip replaced and that slows me down some, but otherwise, I just keep doing what I've always done. I love farming and waking up in the morning and anticipating how the day is going to turn out."

When Robert realized his traditional and aging apple orchards were no longer viable, he turned to turkeys. Although market entry is expensive -$900,000 for quota rights and an additional $600,000 to build a new barn -the supply management system guarantees a set return price, making this in the long term, a smart business decision.

"You see, I had a philosophy of my own. I didn't want to follow a pattern. I always had an interest in agriculture, but I didn't want to farm just because my dad did. Now my son Robert and I farm together. He's a clever fellow and a bit of a visionary who likes to take on new things."

The Chipman farm supported four generations of mixed fruit growers until 1999, when Eugene's son Robert turned his focus to turkeys. "We still grow apples, plums, berries, pears, and have 25 beef cows. But seven years ago I became discouraged with the whole tree fruit industry — I could see that it was no longer profitable. I may be a fourth generation farmer, but I'm a first generation turkey farmer."

Robert's turkeys arrive at the farm when they are only one-day old. The 'poults' are transported within 24 hours of being removed from their incubator and are unloaded by hand into Robert's fully-automated climate-controlled barn. During this brooding stage, the barn is kept at a balmy 35 degrees C., giving time for the young birds to grow feathers that will replace their

down. (Mature turkeys have about 3,500 feathers.) Over the next 12 weeks, the birds will move freely in the barn until they gain a market weight of about six kgs.

In Canada, hormones are not approved for use in poultry production. If antibiotics are required, there is a strict withdrawal period required before the farmer can market his birds. Robert averages seven cycles of 8,000 turkeys each year, for an annual production of 56,000 birds. They are sold to a local co-op who market fresh roaster turkeys.

Robert raises two types of birds, 'Nicholas' and 'Hybrid' hens. There are no Toms in his flocks. Male turkeys are the gobblers while females make a clicking sound. His birds are fed a pellet diet of corn, wheat, soya and bean, with a fat, vitamin and mineral mixture. The young birds grow so quickly that Robert must regularly adjust the feeding ratio to match changing growth stages. He uses 620 tonnes of feed annually.

Canada has the seventh highest turkey production level in the world: the average Canadian eats turkey 15.5 times a year. Turkeys raised in commercial production facilities such as Robert's are grown for specific body characteristics including heavier breasts and more white meat. Robert's roasters end up in grocery stores all throughout the Maritimes.

The extended Chipman family: Gus and Vinnie. It's a rare Canadian farm where a dog doesn't greet you coming up the driveway. Behind the initial warning snarl is usually a teddy bear just waiting for a rub behind the ears. Most play an enthusiastic dual role of guard-dog and children's pet.

The Chipman family, (left to right): Eric, Allie, Emma, Adam, Margie, Robert, Leah and Ryan.

Turkeys are the only poultry breed native to the Western Hemisphere, and it's thought they may have got their name from 'firkee," the Native American word for the birds. Unlike their wild cousins, domesticated birds cannot fly. The heaviest turkey on record was raised in England and weighed 39 kilograms — about the size of a black Labrador.

Turkey production in Canada is protected under a supply management system meant to ensure that consumer demands are met, and that turkey farmers receive a fair profit for their product. In Robert's case, establishing a first-generation turkey farm came with a $1.5 million-price-tag to cover both the quota and turkey barn.

The Chipman farm sits 150 metres from the Annapolis River in the heart of Annapolis Valley. Farming is a dominant industry in the lush valley, which is recognized as one of the

most important fruit growing areas in Canada. Along with its orchards and vineyards, the region is also noted for the highest tides found anywhere in the world.

When Eugene was born in 1913, approximately half of Canada's population farmed. Now, less than two percent of the population are farmers. As small and medium sized farms gradually vanish from the Canadian landscape, we will see fewer father and son teams like the Chipmans. Eugene says he's witnessed phenomenal changes in agriculture over the last 70 years, and he hopes the future will be brighter for family farms. "I'm not retired, but I must look the word up one day to see what it means," he laughs. "I've never quit anything I've tried. Remember, life is a wonderful experience! My advice to others is to keep focused on what you really want to do."

The Riverbend Hutterite Colony

WALDHEIM, SASKATCHEWAN

S AM AND ELIZABETH walk arm in arm through a picture-perfect setting. Identical houses line the main street, each with its well-groomed lawn and pristine garden. The couple smile and nod as they pass young mothers clothed in traditional knee-length dresses with matching 'tuchs' (head coverings). The sun shines in a sky of baby-blue, and barefoot children laugh and play in their yards. This is the Riverbend Hutterite Colony, a tightly-knit farming community.

Built in 1996, the Riverbend Colony is currently populated by 70 people representing 14 families. Riverbend was created when a larger colony outgrew its land base 10 years ago. The colony decided who would stay and who would remain by using an age-old method: pulling names out of a hat.

Sam and Elizabeth Gross were happy to be part of the group selected to relocate. Like pioneers, they embraced the opportunity to create a new life. Sam was one of the founding fathers responsible for selecting the location for the group's new home. "We chose this area because it's great farmland. It's a mixture of green pasture with rich top soil and four kilometres of the North Saskatchewan River runs though it."

Colony members live, farm, and work together communally and each member shares equally in the work and the rewards. The efficiency and success of a team approach is evident not only in their personal dwellings but the ability to profitably farm 12,000 acres. The colony's hierarchy

Sam Gross is the colony's farm boss and carries a huge responsibility for book keeping and supervising all the field work. "But we always have at least some time together in the evenings," says Elizabeth. "With the quietness and a view of the Saskatchewan River flowing almost in our back yard, this is a perfect place to relax."

is made up of a council, voted in by the male members of the community. The council includes two ministers, one farm manager, one field manager, and two council members. The council meets on a daily basis to plan the day, and sort out issues.

Sam is the Colony's farm manager responsible for taking care of the financial books, and planning and supervising all field work. Although a good deal of his time is spent in front of a computer, he still enjoys being out in the field. Combining is what he loves most. "I've always liked farm work. When I was 12, I was given a nice John Deere tractor to drive. I remember driving that shiny green Model A-30 for the first time. Some people might have thought that it was work, but it wasn't for me."

A tour of the common dining facility is impressive. As Elizabeth guides you through the modern scullery and into a central kitchen gleaming with stainless steel commercial cookers and high tech mixers, you know that these women mean business! The eating area has the warm airy feeling of your favourite family restaurant. Seating is boardroom style and easily accommodates 80 people.

Elizabeth says all of the women from ages 17 to 50 take turns doing kitchen duty. "I spend about five hours a day here when it's my turn. It's not easy to feed and clean up after 70 hungry people three times a day, but we manage." Hutterite tradition dictates that during dining, as well as at church, men

The Hutterite schools reflect another era when rural schools were typically one room and contained students of eight grades.
At the Riverbend colony, children normally complete their schooling at age 15, though they have the option to continue with high school if they wish.
The parochial-school system is ideal for the children, none of whom live more than a five-minute walk from home.

Hutterite children are allowed to be children. Until the age of 15 when they are deemed to have reached adulthood, their responsibilities outside the home are minimal. A sense of community is reinforced through play. And the community trampoline remains one of the favourite activities.

sit on one side of the room while women and children sit on the other. In addition to regular meal prep and clean-up, the women make their own bread, butter, jam, preserves and sausage; they freeze their own fruit and vegetables for year-round use. An onsite butcher shop supplies farm fresh poultry, beef, and pork.

Among the members are highly-skilled carpenters, cabinet makers, mechanics, and machinists who have constructed all of the colony's houses, barns and sheds, and furnishings for the homes. Their clothing is hand-sewn, as each woman is well equipped with a dedicated sewing area and the latest machinery. The colony's laundry, wood-working shop, machine shop, bakery, butcher, dining area, and school are all within a stones throw of each other.

The cozy three-room schoolhouse has a current enrollment of 15 children, from kindergarten to teenagers, all who share one classroom. Children speak both German and English, and are provided schooling up until grade 12. During summer months, young people are assigned various jobs in the colony giving them an opportunity to explore their individual areas of interest and to learn a trade. Although the children do not have television or radio, they do have access to home computers, and enjoy lots of freedom to play sports, swim, fish and ride their bikes.

Hutterite colonies are found throughout Saskatchewan, Alberta, and British Columbia. For the most part, the colonies are self-sufficient, growing and raising most of their food requirements. They share a common ancestry with the Amish and Mennonites, but differ in their belief in communal living and the sharing of possessions. All goods are owned communally and all money earned belongs to everyone on the colony. This community-ownership system is based on the Bible, with Acts 2:44-47 as some of their guiding verses: "And all that believed were together, and had all things common...."

Like other colonies throughout the country, Riverbend has adopted modern farming techniques and embraced technology. Tractors are equipped with global positioning systems (GPS) and as Sam says, "We don't believe in staying behind." Members ride horses for cattle herding and branding, and each

Equality is paramount in a communal Hutterite colony. Newlyweds are given a fully-furnished, independent housing unit similar to their parents. The only significant difference might be in size — with larger families occupying up to six bedroom units.

farming operation has its own manager and a few helpers. The colony grows canola, wheat, peas, and barley; they have 500 sows, 400 beef cows, 85,000 turkeys, 14,000 chickens, and a number of ducks and geese for their own use. Their two-acre 'personal garden' fills a commercial-sized freezer and well over 1,000 two-quart jars of fruits and vegetables.

Sam says the colony's challenges are no different than those of other farmers. "Land is expensive here at $800 an acre. All of the smaller farms are gone now — you have to be big to survive. When we bought this land, it belonged to six farmers, but only one was still actively farming. Living here means we are far away from the grain terminal, so freight costs are higher and it costs more to get supplies in. But we do have the advantage of being a mixed farm and having enough labour to get things done."

Sam believes that there should be more education related to farming in the public school system. The colony is very receptive to outsiders and gladly welcomes visitors. "We get bus tours here at the colony and see kids who come from the city who don't know where eggs and milk come from. It is good for children to have an understanding of farm life and where their food originates. Maybe if there was more education, people would be willing to pay more for their food."

Sam, an intelligent, soft-spoken man, speaks of his home and fellow members with great respect. "All the money goes into one pot and the bills are taken care of. If an individual needs something, it is purchased. We work together well and have a good life. When there are issues, we discuss it as a group. When someone gets older or sick, they don't have to worry. There are loving people here to take care of them."

After 30 years of marriage and six children, Sam and Elizabeth are completely at ease in each other's company. When asked who makes the decisions in the family, Sam says, "Well … I guess it's joint." Then he laughs and with a twinkle in his eye, points to Elizabeth, "Or maybe it's her".

In a good year, a single acre of grapes can produce 15,000 glasses of wine.
Canadian wines now rank as some of the worlds finest, surpassing even the traditional Bordeaux favourites.

The Glory of Grapes

NIAGARA-ON-THE-LAKE, ONTARIO

SHOULD YOU PASS THROUGH Ontario's picturesque Niagara Peninsula and see a number of select farms with signs boasting, '20 Bees Proud Grower', don't be confused. These farmers are not in the honey-making business. They are a group of 19 ambitious grape growers — and one wine maker — who have ventured into the wine industry.

The farmers' mission is simple — to produce great wine with their own grapes. The look and feel of the brand is hip and youthful. The wine bottles are marketed with a striking logo featuring a bright yellow number 20 and a farmer holding the reins of a giant flying bumblebee. With that logo and the tagline, "It's a wonderful wine created from the soil to the glass by our growers and winemakers," you know these bottles are not going to sit ignored on the shelf.

Ken Hunter is one of the newer 'bee' inductees. In partnership with his dad, Bob, he grows 75 acres of grapes in some of Canada's most expensive farm real estate. He says, "An acre of land here goes for a minimum of $25,000. We definitely feel the pressure of town moving in on us. In farming, it's an ongoing challenge to keep things viable."

The 46 year-old farmer wasn't always sold on the idea of farming. In his teens he was undecided about walking in his father's farming footsteps. But when Bob told him, "Don't come back until you want to work without a watch on your wrist," the message came through loud and clear. Farming isn't something that happens from 9 to 5. It's a life-long commitment.

Ken's early days in farming were the most financially challenging of his career. "When I began working with my dad, it was very tough to make a living. From 1986 to 1992, we grew peaches and it seemed that everything went wrong. We suffered through low crop prices and bad weather conditions year-in and year-out."

Overwhelmed by financial stress, Ken and Bob put the farm up for sale in 1993. But as luck would have it, they received only one extremely low offer. They decided to try again for *just one more* year. The following year Mother Nature cooperated with a much better crop. The price of peaches went up and everything seemed to fall into place. Finally, they had the year they'd been waiting for! It was the tipping point, and the push Ken needed to try something new. With new-found confidence Ken began to build one success on another.

As the Niagara Peninsula's wine industry had grown tremendously over the last 25 years, Ken decided to focus on grape production. Today, there are approximately 130 wineries in Ontario. In terms of farm gate value, grapes are the second most valuable fruit in Ontario. Niagara vintners continue to demonstrate Canada's success in producing fine wine. At a 2005 wine tasting in Ottawa, expert tasters had an opportunity to evaluate six Bordeaux wines, six Ontario wines, and six wines from British Columbia. One BC wine and one Ontario wine were ranked ahead of the top-rated Bordeaux.

One acre of grapes can produce an average of 15,000 glasses of wine: it takes about two and a half pounds of grapes to produce a bottle of wine. The Greek poet Aristophanes said, "Quickly bring me a beaker of wine, so that I may whet my

The Hunters, (left to right): Ken and Julie, parents Ruth and Bob.

As recently as 40 years ago, grapes were tediously harvested by hand. Rising wages and a shortage of skilled labour forced the evolution of sophisticated mechanical harvesters. Just one of these leviathans can harvest 5,000 bushels a day, replacing the equivalent of about 120 people.

"Having permanently stained hands is just part of the work," says wine-maker Sue Ann Staff, as she brings Ken up to speed on the day's activities. "Wine making is a hectic process, with 70 percent of the years work being done in just two months. I guess one of the benefits of this job is obvious. Someone has to do the testing."

mind and say something clever." Apparently Canadians are whetting their minds at a rate of 17 bottles per year.

Professional wine tasters say that to fully experience the greatness of wine, and get the full benefit from what you are drinking, you must use all of your senses: Your eyes to examine colour; your nose to identify aromas and appreciate the bouquet; your mouth to swish and swirl and take in its flavour.

Ken grows 70 acres of vinifera grapes and five acres of a hybrid. He says that weather has a huge impact on grape qual-ity and harvest. "Like other fruit, grapes are challenged with wet, humid weather and mildew problems. But fortunately, unlike a peach that can't be sold if it has bug marks or is slightly damaged, it doesn't matter what wine grapes look like. They're not a cosmetic fruit.

"Our hybrid is a tough vine so it's planted in the clay soil patches. Our Vinifera is a class of grape that produces both white and red wine. The best time to process grapes is when the sugar level is where the wine maker wants it. If the grapes are disease-free, we try to let them hang longer to develop more sugar. All winemakers make their wine differently.

"We use two different types of grapes for our ice wine. It has to be 10 degrees below freezing before these grapes are picked. That's when they're frozen solid as a rock! The sugar average in regular wine is 18-24 percent, and in ice wine it's 35-38 per-cent. We harvest between December and March."

Last year, Ken made his first grape contribution to the '20 Bees' winery. He says he's enjoying the adventure and that his life as a farmer is one he wouldn't trade. "Our family is lucky — we farm some of the most phenomenal land in the country. But I've learned that you can't farm with just your heart anymore; you have to use your head. And the most important thing is to have fun. If you love your life, you are successful."

Inksou

RICHMOND, QUEBEC

THE JUDGES AGREE. Her legs score 96 points out of a hundred, and her exceptional rump deserves a 97. With an impressive 94 for mammary, she's an absolute shoe-in! No, this isn't the Miss Canada contest. It's the conformation analysis earned by Inksou — a 12-year-old Holstein.

Inksou is a beauty who has had the honour of winning the Royal Winter Fair twice. She is currently the highest-classified cow in Canada with 96 points to her credit. Her owner, David Crack, beams with pride when he talks about her illustrious track record and shows you his office walls, covered in hundreds of ribbons and trophies.

David's door is always open — he loves to welcome people to his farm. Having travelled throughout the world as a cow breeder and auctioneer, he has endless tales to tell. Though retired from auctioneering, David is still a fast talker with an expressive face, big laugh and dancing eyebrows. On the wall to his right hangs a sign: *Live well, laugh often, love much.* It's clear that David does just that.

Until the late 1960's, David's grandfather and father were pure-bred Jersey dairy farmers. But at the age of 22, David brazenly broke with tradition and altered the family's future by bringing home the first Holstein cow. "Back then we were true-blue Jersey farmers. But when the Jersey market started to decline, I felt I had to do something, so I bought a Holstein. When I got her home, I hid her because I didn't want my neighbours to know right away. Soon after, I entered her into the local fair and guess what? She won!"

That bold move was just the beginning. The Crack farm gradually transitioned to 100 percent Holstein. In 1977, they were awarded with a 'Master Breeder Shield' from the Holstein Association of Canada. Now, almost 33

A once-proud chicken barn has fallen from grace to become a simple storage shed for odds and ends. "I should really tear it down," reflects David. "But I love the sight of it. It seems like it just belongs here."

David and Phyllis Crack. Now that he has retired from travelling the world as a sought-after auctioneer, David cherishes the moments when he has an evening free with his wife.

"I've flown all over the world and have been asked to sell cows in many countries — South America, Europe, Japan. I was the first person to send animals over to England and then sell them myself. Back in 1988, I sent 23 head over at an expense of $4,000 for each cow. It was a huge risk. I had no idea how buyers would react as it was something completely new. After three months of having the animals quarantined, it was so exciting to see *my* Holsteins come down from the barn into the auction ring. The crowd was drawn to them right away. And at that second, I knew they were going to sell!"

As is often the case in the world of marketing, perceived value can be more important than actual value. "One time at an auction in Italy, one of the Pope's heifers came up for sale. It was a two-and-a-half year old who hadn't yet calved, and that normally isn't a great sign. An animal like that might go for around $1,000. But because this one came from the Vatican, it went for $8,000. You just never know."

The success of the Crack farm today depends not only on dairy farming and on auctioneering, but on the complex science of breeding — and on the donor embryo business. To date, 200 of Inksou's embryos have been sold to places as far away as Japan and Europe. Future embryos are currently on a

years later, they raise 140 Holsteins on 250 acres of rolling hills in Richmond, Quebec.

David has long been a professional judge of show cows. As a young farmer he attended several judging schools and has since made his mark with a profitable track record. A star like Inksou is judged on many factors. First, she needs to be big — around 800 kgs. She must be a good milk producer. Judges look at longevity, legs, and a good appetite. The cows should be big ribbed with a long body, implying a big capacity for roughage: the more roughage they eat, the healthier they are. Finally, the body should blend well with a symmetrical udder.

In a world where one animal can command hundreds of thousands of dollars, a skilled auctioneer like David plays a critical role. The most expensive animal David has ever auctioned off fetched an astounding $500,000. It's exciting work that offers top auctioneers the opportunity to travel extensively and create friendships around the globe. David describes auctioneering as a performance art. "You have to bring folks into what you are doing. You have to work at getting people involved and making it exciting. It's about creating action.

Reminiscent perhaps of an era when doctors made house calls, veterinarian Raymond Houde visits the Crack farm weekly in his mobile lab. In addition to keeping the Crack's pedigree Holstein herd in prime health, Dr. Houde identifies, extracts and freezes donor embryos — most of which are shipped to Japan. His annual vet bill? About $70,000!

David Crack is justly proud of Inksou. She is the highest ranked show cow in Canada — and was valued over $100,000 in her prime.

pre-order contract to a buyer in Japan for an impressive $900 each.

Inksou's genetic heritage has been carefully managed. With documentation going back five generations of pedigree cows, and with her offspring worth $25,000 each, she is the kind of prize cow breeders aim for. According to David, superior genetics is about creating an animal that gives a lot of milk and is easy to work with — the type of cow every farmer wants. Breeding science has produced Super Ovulation, a veterinarian controlled process which involves injecting the animal with hormones. When there are 15 to 20 eggs in the ovaries, the embryos are removed, frozen, sold, and shipped. Transplanting might be six months away.

Breeding is a demanding and expensive business. Along with the high medical costs, there are health issues, rising input expenses, and failures: the breed crosses that just don't work. It isn't uncommon for David to receive vet bills of $4,000 to $5,000 at a time. But David's dedication to providing a high level of animal care is a key reason for his success. "Animal welfare is the number one priority here. We provide vitamins, great feed and bedding. The more I do for my cows, the more they do for me. On our farm, the cows come first."

The future for the Crack farm looks rosy. Phyllis, David's wife, has just retired from her role as auction bookkeeper. David and his son, David junior, are working the ten-year plan. "David junior is a cow man like me — he loves good cows and judging. And as he gradually takes over, Phyllis and I will have more freedom. But I think I'll have to practice and maybe slowly train myself for retirement. I guess I'm just not a veranda-sitting kind of guy!"

Failure is not an Option

BEDEQUE, PRINCE EDWARD ISLAND

NATHAN BURNS IS A CHANGE MAKER. At the age of 25, he bought out his father's poultry operation and learned to embrace risk. In the five years that followed, he expanded buildings, added cages, installed a refrigeration unit, purchased 8,400 additional layer hens, married his girlfriend Wendy and became dad to three kids.

Nathan farms 90 acres on Prince Edward Island where he grows milling wheat and raises Shaver White Hens for egg production. The confident decision-maker is a straight forward man who looks you right in the eye when he speaks. "The days of grandma heading out to the chicken barn with a basket and pail of grain are long gone. If I farmed the way my dad did twenty years ago, Wendy and I wouldn't be able to survive."

Nathan's strength as a farmer is his ability to look at an opportunity, evaluate the risks, and then execute a plan with unshakable resolution. "People tend to fear and resist change, even when it's good. I don't. One of the first things I did after buying the farm was to purchase a $50,000 state-of-the-art egg packer. And right now we are building a whole new facility, one that will increase production by 38 percent without hiring more labour. It is a huge financial risk, but a calculated one."

Nathan's new chicken barn will house 11,500 of his total of 30,000 Shaver Whites. These hens are fairly lean, calm birds that produce white eggs. Their feed and water is carefully monitored using a computerized system. And as Nathan says, "If everything is great with my hens, they will lay eggs."

His farm is considered above average size in terms of the number of chickens, but there are also massive egg farms that house hundreds of thousands of birds. Farmers in

With a daily production of 30,000 fresh eggs, the Burns family have no qualms about scrambling just 17 of them for a family lunch. Four-year-old Mallory prefers hers sunny-side-up, thank you.

Canada care for a total of about 26 million hens, with the egg industry contributing $410 million annually to the economy.

Most chickens in Canada are raised indoors in cages with precise environmental controls that mechanically dispense water and feed. There is no difference between brown and white-shelled eggs: colour is determined by the breed of the hen. Chickens are considered average when it comes to the size of their eggs. Hummingbirds lay the smallest eggs and ostriches lay the largest. The heaviest egg on record weighed about the size of a small sack of potatoes, 2.35 kgs (5lb 2oz), and was laid in 1997 at an ostrich farm in China.

Research and technology has led to superior birds, and major advancements in nutrition and management. In 1960, the average hen laid 193 eggs per year, compared to today's average of 300 eggs. From Omega-3, to vegetarian eggs (hens fed a diet of 100 percent plant origin), to free range eggs, consumers now have greater choice, and more influence on the care, feeding, and management of the hens producing eggs.

Nathan follows the 'Start Clean — Stay Clean' program. This is a national program designed to ensure the production of clean, safe, quality eggs. The program has standards that must be adhered to by farmers, and egg barns are regularly inspected.

Nathan Burns is no stranger to hard work or to progress. "I try to improve our operation, somehow, every day. This is a changing world and I want to keep up with it."

Nathan discovered an opportunity to sell his product to the industrial market as opposed to the fresh egg market. His hens produce approximately nine million eggs annually; these are sold for commercial products like salad dressing and baking products. His eggs, like those bought at a grocery store, are not fertilized (a hen doesn't need a rooster to lay eggs) and will not hatch.

*The Burns family, (top to bottom): Nathan, Wendy, Mallory, Kendall and Mitchell. On a blistering July day, the ultimate cool-down is just a few minutes inside the 10 degree (C) egg cooler. Even more than cooling, the most effective way to extend an egg's shelf life is **not** to wash it. A just-laid egg is coated in natural oils which effectively seal the shell, protecting its freshness.*

They are not quite designer chickens but, by varying the feed ration, Nathan's chickens will produce eggs with yolks matching any one of the 15 colour choices. The process has been developed to meet the capricious and changing demands of consumers.

Managing an operation like Nathans requires paying close attention to detail. "The farmer needs to have an understanding of nutrition and what the consumer wants from the end product," says Nathan. "The best birds with the best genetics are only part of the equation. Management is crucial. The perfect egg is nicely oval with a smooth shell and even unwashed it's very white with a dark yolk.

"In the future, I see more people eating eggs that are not in their natural shells — they will be processed and in a container. The trend is going away from fresh. But if consumers want eggs in a carton, that's what they're going to get. Our family still eats lots of eggs. Most days we go for scrambled."

His wife, Wendy, is a busy stay-at-home mom with three children all under the age of five. She too was raised on a farm, and the pair have forged a solid partnership of farm and family. "I couldn't do this without Wendy. It's a team effort — we make decisions together. She looks after the financial books, the kids, and she looks after me!

"As far as the chicken operation, I know things will continue to evolve. In farming today, we need to know how to assess change and adapt to it. It's simple. If what you're doing today doesn't work, then why would you keep doing it? I'd like to try growing organic garlic next year and maybe alfalfa for high quality horse feed. I know I'll succeed because if you present me with a problem, I'm determined to find a solution. Failure is not an option."

Hard Core Aggie
A young farmer's perspective…

TEHKUMMAH, ONTARIO

THE MAN: "I don't want an office job. I need to be out-side in the fresh air. And I do not want to work for any-one else. That's why I'm a farmer! I grew up on a farm and the only time I've ever lived in a city was my four years at the University of Guelph getting my Ag degree. But that was enough — it's the rural life for me! When I was little, I spent tons of time with my dad when he was working. It's a bonus being a farmer's kid.

"I'm outdoors a lot of the time with my dog Tippy. I enjoy biking, playing the guitar and drums. No, I don't have a girl-friend, but I'm looking, and someday I hope to marry and have a family.

Alex Anstice is 22 years young, with a degree in Agriculture that is so fresh the ink is still wet. While an undergraduate at the University of Guelph, he and a team of students won the North American Intercollegiate Dairy Challenge. Alex is intelligent, confident, and defi-nitely up for the challenge of one day taking over his dad's dairy farm.

Alex is confident any maple walnut ice-cream cone purchased on Manitoulin Island couldn't taste any better. Milk from his farm supplies the local dairy, well-known for its gourmet ice-cream made with real cream and real butter.

Three generations of Anstice dairy farmers, (left to right): Alex, Ron and Jim.
Appropriate to the area's geography, the farm name is derived from the local Ojibwe native language and means "Rolling Hills."

The place: "I'm a 'Haweater'. That's what you call some-one who was born on Manitoulin Island. It's a small place. We're two hours away from the closest mall and we just recently got a bowling alley. It's not densely populated and there are lots of lakes and bush trails for biking. That's the best part — the space, the water, and clean air."

Manitoulin Island is the world's largest freshwater-lake island. Year-round access by motor vehicles is permitted by a swing bridge that crosses the North Channel. The soil is relatively alkaline, which pro-motes the growth of blueberries and the famous 'hawberry' — the berry that 'Haweaters' eat. The island has a year-round population of 14,000.

His history: "I remember falling asleep on the back of my dad's tractor when I was little. By nine years-old I was driving the tractor myself and by 12, I was helping with the milking. My dad never told me I had to work — I took the initiative. I'd carry grain to the heifers every day and help him and my Grandpa fix fences and repair stuff."

Alex is the son of Ontario dairy farmers, Dorothy and Jim Anstice, and grandson of Ron and Justeen Anstice. All family members live and work on 275 acres where they grow alfalfa, hay, corn and barley.

The job: "Each morning, at six o'clock, I roll out of bed right into the barn to milk our 40 Holsteins. Our cows are quiet, gentle animals with their own distinct personalities. The first thing I do is chase them into the holding area. You can tell the best behaved ones because they move into formation right away. Then I load up the milking parlour with five cows on each side. We do rotating sides of five, so it takes seven trans-fers to milk our cows. After that, I'm ready for a big breakfast at my grandparents' house. And I spend the rest of day doing what ever has to be done, like field work or fixing machinery, before I do another milking at five."

Economists predict that small farms in the next 15 years will con-tinue to go out of business unless they expand or specialize. Dairy farm-ing is a rigorous full-time job with no days off. Cows need to be milked when their udders are full — twice a day, every day

(Left to right):
Ron, and son, Jim.
A distinct advantage of a
multi-generational business
is the opportunity to pass on
knowledge first-hand.

The challenges: "Of course, the main challenge is the cost of the quota. It would be impossible for me to milk cows if my family wasn't already into it. I have a friend I went to Ag University with, who would love the opportunity to dairy farm. But his family are cash croppers, and he could never afford to do what I'm doing."

Being a successful dairy farmer today requires knowledge of animal biology, genetics, nutrition, animal behavior and welfare. Dairy farmers must be mechanics, knowledgeable about feed crops, environmentally responsible, and have strong business skills in accounting and technology.

Vision for the future: "It's interesting. Our industry is dependent on new developments in science, technology and economics. And to be profitable, we need to be educated in all of these areas. Everyone in the world eats and someone needs to provide for them. But there's a lot more to farming than people think.

"Public perception of where food comes from is becoming a huge factor. Farmers are required to change the way we farm based on consumer preferences — like producing organic food, or changing the milk composition to a higher or lower fat. I know a lot of young people are going towards being vegetarian and I think that will be more of a trend in the future.

"I'd like people my age to know that we do our best to keep animal welfare at an optimum level. It seems that the public is naïve when it comes to farming — they pay attention to the negative stuff, even when they might not know the whole

Tippy agrees with her owners. Drink Milk.

story. It's not like in the States. It's very different there — everything is scaled up with big corporate farms. I want people to think critically and make sure they get the whole story."

In Canada, organic farms make up less than 1.5 percent of farms today but are the fasted growing agricultural sector — by percentage.

The number of young people in farming is going down — only 11.5 percent are under the age of 35. The value of milk from Ontario's dairy farms is approximately $1.4 billion annually and accounts for about 19 percent of all agricultural production.

Alex's plan: "A farm consultant looked at our operation — situation, assets, liabilities — and helped me put together a plan to acquire 10 cows this year and 10 cows next year. These 20 cows will fill the barn to make it more efficient. Right now, it's only at 60 percent capacity, but the opportunity for growth is already here. We have the facilities and the equipment. The workload won't be much more but I'll need to take out a heck of a loan. Hopefully, the system is stable enough to pay the debt back within 10 years. So during that time, we won't make extra money from having the additional quota. You know, it's kind of scary!"

The average dairy farmer in Ontario is 47 years old with a herd of 54 milking cows. Expanding a dairy herd may require servicing a large amount of debt. The cost to purchase one cow for the purpose of selling milk in Ontario is $30,000. Alex's plan will cost approximately $600,000.

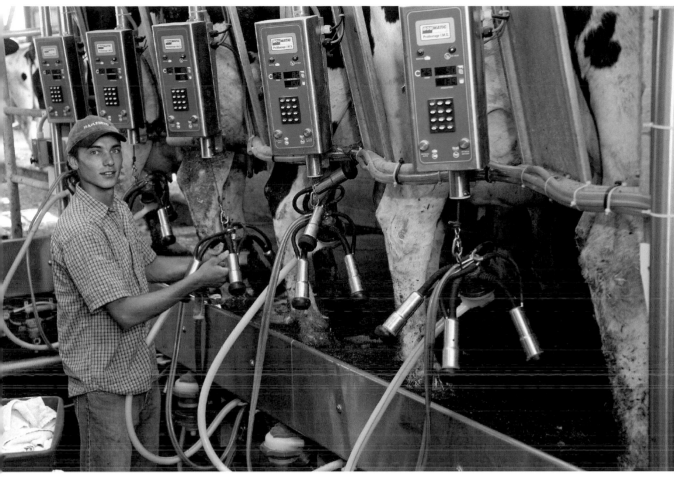

"I got my university degree not just for the piece of paper, but for the scientific knowledge to make me a better farmer" says Alex. "Agriculture is changing, it's on the move, and I want to be part of the change."

Hard Core Aggie: "I shared a house with a couple of other 'Aggies' — that's what they called us students in the Ag program at the university. I was in the hard core group. I didn't wear what they wore — cowboy boots and hats. They were just real proud to show off their farming background. In the first year, some of the Aggies set three piglets loose in the south residence. It was a classic Aggie prank — each pig had a letter on it, O, A, C, — for Ontario Agriculture College.

"School was an opportunity to meet so many different kinds of people and make lots of friends. It's a great place for a young farmer like me to build networks with classmates and tap into resources that I can use in the future. The Ag community in Ontario is very tight-knit.

"I didn't get a degree just so I'd have a piece of paper. The purpose of university is to teach you how to learn, how to solve problems and tie concepts together. And these are the skills that are the most valuable in my day-to-day life. An example is our feeding management program. Managing a cow's nutrition just before she calves is very important to help ward off diseases. When I returned from school, I altered the pH levels in their feed and this resulted in a reduction in the number of cows coming down with milk fever. I'm using my degree and scientific knowledge everyday on the job. Our industry is always changing, and the Ag program taught me how to stay up-to-date with what's going on.

"One thing I learned from my dad is that even with all the science and all the studies done on agricultural topics, there is still a certain amount of instinct involved in farming. Life experience with the weather, animals and crops, can never be replaced with science. On the farm right now, my dad leads, and I'm right in the middle between him and my grandfather. But over time, I'll be the one making the managerial decisions."

"I get to go to work on my horse," says Gary. "Riding for me is about connecting humans and animals.
Twist, my favourite horse, and I easily understand each other's moods."

Misatimawenaw
The Horse Man

PAYNTON, SASKATCHEWAN

GARY PEWAPSCONIAS GALLOPS FULL SPEED through a rock-strewn, rolling pasture. There is no question of who is in charge. Twenty geldings respond instantly to his shouts, "Hip! Hip!" The horses bolt in a new direction, cutting a clear path to avoid a worried herd of cows and their confused young calves.

Gary is an extraordinary horseman who excels at the training and buying and selling of horses. At 5 feet, 11 inches, he sits tall in the saddle on Twist, his muscular, chestnut brown quarter horse. Gary rides with a natural grace and elegance that comes from decades of experience, and from generations of knowledge passed on by the elders of the Plains Cree Nation. His gentle nature perfectly matches his passion for horses. "My father and grandfather were both horsemen. I've been riding since I was five years old," he says.

It can take up to a year to train a three-year-old horse who has never been on a rope. As Gary attests, when things go wrong, both trainer and animal can get hurt. "I've broken my left collar bone three times and have been kicked and stepped on lots."

"We are meant to be connected to the earth. Riding is one way of slowing down and appreciating where you are."

In the first years of their marriage, Celine and Gary travelled the rodeo circuit — carefree days and night-time camping with the biggest question being, "Will it rain tonight?" Twenty years and four children later, they have still retained their sense of play.

The first step in training is to 'tie break' the horse. Under Gary's watchful eye, the young animal is roped and haltered to a post. Depending on the horse's personality, the tie-break can take anywhere from six to 24 hours. Once the animal has gained respect for the halter, the next step is to 'lead break'. Gary begins this process by slowly pacing the horse with a lead, all the while talking to him softly and stroking him to build trust.

The final step is to hook the horse up to a wagon. Gary calls this *taking the bronc out of them*. "When I buy a horse, I have to determine if he's had bad experiences in the past that might cause a problem. It's something I can tell by looking at him. Horses are all different, just like people. I can read their disposition by looking at the shape of the head, the thickness of the ears, the nose, the jaw, and the way they walk. Soft eyes can tell you a lot about a horse's personality," he says. His remarkable intuition is one reason Gary is so successful with animals, and is well-respected by people in the horse business.

Gary grew up with 12 siblings in a family where tradition and culture were important. He remembers simpler times when people lived closer to the land. "I grew up being connected with animals, plants, and the rain. Native people have a love for the land. We didn't have electricity. Actually, I was 13-years-old before I turned on a light switch for the first time. Although we didn't have things like radio or television, our people entertained each other by sharing our history through story-telling.

"My grandfather, Adam Twotwosis, had many stories, but this is my favourite: When he was a young man, Adam lived in a camp where there were many wild horses that no one could catch or tame. One evening, he went out alone and tracked a herd of horses for nearly 20 miles. He followed them for several hours and when the horses tired, he steered them east towards his home. By late evening he led them right into the camp. Over the next few weeks, he worked them and eventually tamed them. It was an amazing feat!"

Gary's riding skills, particularly as chuck wagon outrider, have taken him to centre ring at the Calgary Stampede. Advancing to the Stampede's World Finals in 1979 won him a prize buckle.

Gary, his wife Celene, and four children live on the Blue Hill Cattle Ranch. When Gary and Celene were newlyweds they traveled the rodeo circuit; they have many wonderful memories of being young and carefree. "We camped out every night," Gary says. "The only worry we had back then was whether or not it was going to rain."

In those early days, Gary was a skilled chuck wagon outrider and rode in the 1979 Calgary Stampede. Horse skill and competition is still a family affair for the Pewapsconias. From the youngest to oldest, they all compete in Team Cattle Penning, an action-packed, arena sport. Riders on horseback have 75 seconds to round up numbered cattle. The team, made up of three riders, who successfully pens the most cattle in the shortest time, wins. Judging by their ribbons, trophies, and shiny buckles, the Pewapsconias family has won more than their share of races.

The Blue Hill Cattle Ranch that Gary manages is owned by the Little Pine Reserve. The Reserve is a community-owned land base that offers families who live there 'the right of use.' The Pewapsconias family invests in stock, works the land, and makes improvements, but they also dream about someday owning land in their family's name. They currently own 45 cattle and 80 horses but Gary says, "Government programs are limited. There just isn't anything out there to support the horse rancher. Everything that we own, we got by going without and working hard."

Along with helping to develop legislation and head up special projects for the Reserve, Celene volunteers with 4-H Clubs teaching children horsemanship. She's a strong, confident woman who is at ease sharing her philosophy of life. "We are fortunate to have so much when there are so many families who have very little. Teaching children to work with horses helps to build self respect. And only by having respect for themselves will they learn to have respect for others.

"I believe there is a hierarchy and a natural way of the universe: Mother earth is first. Second is plant life and small life forms. Third is water and sky life, then there are the two and four-legged animals. At the bottom of the hierarchy is humanity. In our society, we people are acting like we are at the top. This is wrong. We must learn to have respect and love for the land."

To qualify as organic, cabbages raised on the Gillam farm must remain chemical free. The consequence is labour-intensive hand weeding.
"My 17-year-old daughter Kara is the most motivated employee I've ever had," says her proud father Elvis, "And has no trouble keeping up with the men."

JIGGS DINNER

ST. FINTAN'S, NEWFOUNDLAND

ASK A HOMESICK NEWFOUNDLANDER about *Jiggs dinner* and he'll tell you fondly about a traditional east-coast dish with a very distinctive smell. The boiled dinner, no doubt a staple in his mom's kitchen, consists of salt beef, turnip, cabbage, carrots, and the old standby — potatoes. Guaranteed to warm your insides on a cold Atlantic day, these hearty vegetables, like the people who live on *The Rock*, thrive in the windy, salty climate.

As a young boy growing up in a rural area, Elvis Gillam fondly remembers his mom's Jiggs dinners. Now a farmer committed to growing the best root vegetables possible, Elvis supplies all of the necessary Jiggs ingredients for others to enjoy. "In Newfoundland, we feel that we grow a superior turnip to anyone in Canada. Our cool weather conditions keep the sugar content high and the turnips sweet and tender. I've got lots of perfect turnips that weigh in around two and a half pounds. I also grow beets, carrots, cabbages, and spuds. I have a few specialty crops, but where we live, customers definitely prefer the more traditional vegetables."

Elvis and his wife Marilyn, farm in St. Fintan's — part of a clustered community of 700 people located about a kilometre from the ocean. Their farm was passed down from Elvis' great grandfather who paid $29.20 for a land-grant block of 98 acres by the river. Although his dad was a construction worker, Elvis was attracted to farming. "As a kid, I joined a 4-H club and my very first project was cooking with a group of girls. I loved it. The next project was gardening and that's when I got hooked on growing things.

"I guess you could call me a workaholic because work is my life. We opened a farm market this year so, between field work and overseeing the business, I am running constantly. Selling direct to the public has certainly put us through a learning curve. We're finding our customers want us to be everything to everybody. They want the convenience of a one-stop shop. And we are doing our best. Next year, we have plans to open our own in-store market bakery.

"People buy from us because they trust us and know that we're growing our vegetables as naturally as we possibly can. If I say we don't spray our broccoli and cauliflower, we don't. And people know that I'm good to my word. One day, while I was working at the market, a customer approached me who was quite upset. She had just seen a green bug crawling on one of our cabbages. I told her straight, Well, you can always go somewhere else and buy a chemical-sprayed cabbage that doesn't have bugs. It's your choice!"

Less than one percent of Newfoundland's land is classified as farmland, supporting 750 farms. With a population of less than half a million spread over an area of 111,390 square kilometres, product distribution is one of Elvis' biggest challenges. "This is a big province. We farmers in Newfoundland are still in our infancy. We lack organization, and co-operation between ourselves. It's a huge challenge to get my crops to other parts of the province.

Building a market garden in nearby Corner Brook has been a challenge for the Gillams. People traditionally have not been willing to pay a premium for organic vegetables. At only $6.99 a bag, their potatoes might be a loss leader, but a good marketing strategy to build a clientele.

Elvis and Marilyn Gillam both grew up eating at least one Jiggs dinner a week. "But it's something you never tire of," Marilyn waxes enthusiastically. "The vegetables are organic and fresh right out of our garden. Try matching that in any restaurant."

"The winters here can be nasty but I have seen our weather patterns change dramatically over the last 15 years. It seems that springs come quicker, and each year is a little milder. This year, we had a good, warm, early spring but then it started raining. And it's been rainy ever since! The fields are saturated and the humidity brings more bugs. We are really hoping for a long Indian summer."

The Gillam farm offers family members the opportunity to wear many hats. From truck driver to manager, field hand to marketer, they seem to be up for the challenge. "We work well as a family and when we are together for a meal, our dinner table becomes the boardroom. Marilyn drives the produce to the market daily and oversees advertising, stock inventory, and staff. Our 17 year-old daughter Kara is a terrific worker. She never complains. When she's not in school, she works at the market or helps in the field seeding and pulling weeds. For me, there isn't one part of my job that I don't enjoy. I like it all!"

Harvard Farmer

CHATHAM, ONTARIO

LIKE MOST CHILDREN raised on a farm, Bob Kerr was given a special gift at an early age: the gift of learning by *doing*. "As a small boy living on a farm, my environment was the open land. I had the freedom to explore creeks, catch tadpoles, and get my feet wet. Childhood was a wonderful time. As long as I was back by dinner time, no one worried. That was when I first began to challenge myself and acquire good judgement. I had a stimulating and challenging formal education.

But what you learn year-to-year is more important than any academic qualification."

Bob is a Harvard Business School graduate who attended the prestigious Cambridge school in the late sixties. He now raises natural beef and farms organic vegetables, commodity crops, and *lots* of tomatoes. His 1,700-acre farm located near Chatham, Ontario, is on the sandy loam soils of the Thames River Delta. Rich in calcium and minerals, the Delta provides

"I'm fortunate to have good help," says Bob. "Mafalda, for example, has been with me since the seventies. She loves the seasonal work and being part of the harvesting."

good workable land for serious tomato growing. Bob produces an amazing 11,000 tonnes of tomatoes annually.

Three percent of every bottle of Heinz Ketchup is made up of Kerr's tomatoes. Ketchup was introduced in 1876 as a "Blessed relief for Mother and the other women in the household!" Heinz buys more tomatoes than any other company in the world—over two million tonnes each year. Heinz Ketchup can be found in 80 percent of the fridges in Canada. It comes as no surprise that the biggest consumers of ketchup are kids, ages six to twelve.

Bob has been on contract with Heinz since 1975. He makes the distinction clear: these tomatoes are *not* grown for the consumer market, but for the corporate market. The corporation dictates both the variety of tomatoes he will grow and the number of acres he will plant.

Each year, Bob receives over 2.5 million tiny tomato transplants from local greenhouse growers who in turn received the original seeds from Heinz. Planting starts in May and harvest begins in mid-August. According to Bob, efficiency and timing are everything. "We do our own trucking direct to the factory. We have 27 trucks going out from our fields weekly and each truck carries a 40-ton load. In this particular tomato business, acidity, flavour, consistency and yield are important. Colour is not. We're paid about four cents per pound. A good crop is when tomatoes cover the ground. A great crop is when they cover the ground two layers deep!

"Our certified organic tomatoes are in a different category. We started to grow this niche market product to escape the commodity system when we realized that the way we were farming wasn't working. We were being squeezed: we received less profit, yet were expected to produce more product. We grow the organics for the Thomas Canning Company's Utopia Brand. The tomatoes have to be a good size, good colour, and firm with no blemishes. And they must be bursting with flavour!"

During a good harvest, as many as seven of these
40 ton trailers roll out daily from the Kerr tomato fields.
The smattering of yellow renegades will simply be absorbed into
the great ketchup machine. The Kerr's annual tomato production of
11,000 tons would fill approximately six million bottles of ketchup.

Bob also raises 180 Angus beef cattle in a less-controlled and more natural environment than most farms offer. "We've acquired a deep conviction about our animals. They live their whole life on pasture—it's a healthy and happy life that is not possible in a confined feedlot pen. I feel that we are respecting and honouring these co-inhabitants of the earth and not just exploiting them," he says. "We learned from experimenting that meat from animals that live free range in their natural environment, and eat a natural diet, is indescribably delicious. We market to customers who are looking for a product raised without the use of growth hormones and antibiotics."

Bob says the pace of change in farming has been overwhelming and that the future for agriculture doesn't look bright. "We are on a treadmill. We race to maintain our position as the lowest-cost-producer, and there is no bottom. I don't think there's a magic wand out there that's going to give us the answers we need.

"When I took over from my dad, it felt much easier. My career is upside down right now. I put in 10-12 hour days, sometimes seven days a week, because I'm tied to the business that I started.

Bob and Moira Kerr. "With tomato fields all around us, you would think we'd get tired of eating them," says Moira. "But a just-picked, organic, beefsteak tomato is still one of my summer favourites."

Organic and conventional farming are often viewed as diametrically opposed. Bob Kerr has one foot in each camp by growing conventional tomatoes for the ketchup market and organic for consumer use. "There's a place for both markets. Some people focus on what's cheaper, others on quality and taste. I think the day will come when we'll all grow organic."

Now, profitability is the lowest we have ever seen. Canadian farmers need the safety net that U.S. farmers are getting. We're in a global competition about who can produce the cheapest food. That only further fuels the race to bottom.

"My dad and mom had two children, a boy and girl, and they expected me to be the successor. I bought into that expectation. We farmers live intimately in the natural world. Farming is about getting dirt under your fingernails. And what we gain by that intimacy is an appreciation for the beauty and complexity of nature."

"When I'm with these baby chicks it feels like I'm cuddling with my favourite blanket," says six-year-old Benjamin. These 180,000 chicks were hatched in the morning, and delivered to their new Thiessen broiler home that same afternoon. In a mere 40 days, they will be ready for market.

The Land of Opportunity

ABBOTSFORD, B.C.

JOHN THIESSEN WAS IN HIS EARLY 20s when he immigrated to Canada in 1949. The young Russian Mennonite, who spoke very little English, disembarked from the deck of the ocean liner *S.S. Samaria* with just one piece of luggage: a small wooden box containing all of his worldly goods. Soon after, he was reunited with his bride-to-be, Lena. John and Lena happily settled in Vancouver, B.C. Through the years, they scrimped, saved, and worked together and were finally able to purchase a small family farm in Abbotsford. This was the beginning of a thriving poultry operation. During this time, they also raised three sons — a police officer, a teacher, and a farmer.

The youngest son, Rick, is a farmer with the Midas touch. Rick has turned his parent's modest poultry operation into a $17-million dollar success story with 110 acres of prime B.C. land, 180,000 broiler birds, and a trucking business. Rick was named the Outstanding Young Farmer for 2004, and is currently the President of the B.C. Chicken Growers Association.

For Rick, life is a race. He's spread thin between running the business, lobbying the government on behalf of chicken producers, and being an active dad to four children under the age of 10. Rick and his wife Karen, live on the outskirts of town in a beautiful, spacious home cradled in a lush, green landscape overshadowed by mountain peaks.

Rick and Karen met in 1992, and it was a case of 'like at first sight.' "We were introduced at a Halloween costume party at a friend's house who was trying to set us up," recalls Karen. "I was dressed as a hippie, and Rick made a pretty good Garth from Wayne's World. The first thing I noticed was his smile. He hasn't changed much. He's a pretty restless farmer who's always busy looking for the next challenge."

The allure of farming has already caught the imagination of their son Brandon. "Since day one, that boy has loved anything to do with farming," says Rick. "He would be in the barn all day long if we let him. His bedroom floor is covered with tractors, barns, and farm animal toys. Even his walls are painted with a mural of our barn." Rick says the youngster's fascination with all things farming was a prophecy announced at his birth. "It's funny, the first thing my dad said when he laid eyes on him was, 'Now I know why we bought the farm!' Dad couldn't have been more right."

In the broiler business, there's no debate about who came first — *it's the egg before the chicken*. Rick's production cycle begins with eggs, specifically, broiler hatching eggs. Broiler refers to a chicken raised for meat — an entirely different breed from egg-laying chickens. Incubator eggs take 21 days to hatch. Rick receives 180,000 chicks at a time, direct from the hatchery, via a heated truck. When they arrive, his broiler barn floor turns into a sea of tiny, peeping yellow birds.

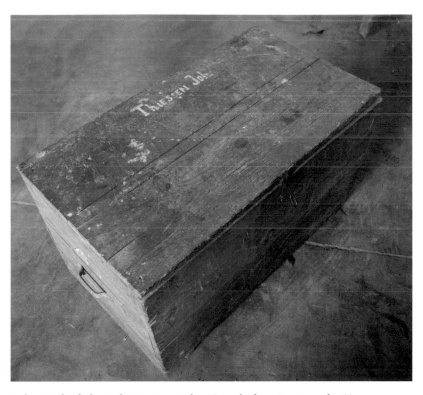

When Rick's father John immigrated to Canada from Russia only 60 years ago, everything he owned fit into this self-made travel box. Just one generation later, the Thiessen agri-business encompasses eight broiler barns and three hog barns, employs six people, and is valued over $17 million.

The Thiessen family, (left to right): Emily, Rick, Ryan, Benjamin, Brandon and Karen.

Like father, like son. Nine-year-old Brandon already has a heart for animals and farm life.
"I like pretending I'm a real farmer. Someday, I want to be a veterinarian."

The fluffy soft chicks are placed on sawdust-covered floor and kept at a cozy 30-32 degrees. The chicks have the freedom to roam the entire barn with open access to clean water and grain — just the right conditions to grow into what Rick calls a *perfect flock*.

"A great bird is about five pounds (2.3 kg) and fully feathered, snow white, clean and plump. The perfect flock is even in size, has a low mortality rate, and converts well from feed to meat."

Canada has a Code of Practices for the Care and Handling of Poultry, which is a strict set of guidelines that all chicken producers must follow. The code is reviewed and revised in cooperation with the Canadian Federation of Humane Societies. The history of the B.C. production of broilers as a meat bird began in the 1920s when producers began exporting to Europe. In 1961, when production reached a new high, farmers formed a Broiler Board as a method of controlling production and stabilizing prices. It was the first chicken board in Canada.

Rick doesn't plan to slow down anytime soon. The high-energy, 37-year-old has turned his eye to the future. "When I was four years old, I remember finding an egg that didn't have a shell — just a membrane. I was fascinated with it. And I'm still fascinated with farming and I love the challenge of it. In the future, we'll protect ourselves from risk by not having, pardon the pun, all our eggs in one basket. We won't be going bigger in broiler chicken commodity, but I'll definitely be looking at other opportunities in some other part of agriculture.

"My dad passed away in 1997 from cancer. He had a couple of sayings that I try to live by. Live and let live. And be happy with what you have."

40,000 Strong!

ST. MALACHIE, QUEBEC

ON DECEMBER 10, 1993, a young chicken farmer from St. Malachie, Quebec took a bold political stand. In a crowded, Chateau Frontenac conference room, 600 hundred journalists, delegates, and farmers watched closely as Stewart Humphrey approached the minister of agriculture, Ralph Goodale. In his left hand, Stewart carried a powerful symbol — the key to his farm.

Stewart approached the microphone, extended his hand to the minister and said, "Mr. Goodale, please take this key with you to Geneva. Let it be a reminder that you carry with you the future of our farms." That brief exchange sent a clear message from the top leaders of a Quebec farmers union to the negotiators at World Trade Commission: "Canadian farmers deserve fair representation at these talks. *Our* future is in *your* hands."

According to Stewart, the UPA, *L'Union des producteurs agricoles,* with its 40,000 members is so powerful that the Prime Minister appoints his Minister of Agriculture, only *after* he makes a call to the union's president. Stewart is on the regional chicken marketing board and a member of the local farming union.

'Ross', a privately-developed strain, is the most common chicken variety raised on Canadian farms. In just 35 to 40 days from hatching, they are ready for market, weighing an average of 1.7 to 2.5 kilograms. Chickens have the most efficient weight gain ratio of any meat product. Producing one kg of chicken meat requires an average of 1.7 kg of feed. On the least efficient end of this same spectrum, one kg of beef requires six or seven kg of feed.

"I inherited my passion for both the union and chicken farming from my dad. I think the union has proved to be well worth the money," says Stewart. His father and grandfather became active union men during the depression era. "When my grandfather and dad farmed, farmers were very poor and had no power. Their families were starving because farmers couldn't get a fair price for their crops or their animals. My grandfather started doing union work in the 1920's and in the 1950's, my dad helped convince farmers to become organized, and to get involved in milk, poultry, and egg boards. He visited their farms to enroll them, and sold them union cards for $2.00."

Stewart farms 250 acres of wheat, hay and grain and produces 220 gallons of maple syrup each year. He also operates a poultry operation with 29,000 broiler chickens sold throughout Quebec and Ontario. Stewart describes the poultry business as solely focused on production: "You don't become attached to animals in this type of farming because success comes from producing the largest volume of meat in the shortest turn-around period possible."

Stewart takes pride in maintaining high environmental standards: he says the public needs to understand that farmers like him work hard to be accountable. He has qualified for the bronze category from Merite Agricole. This provincial agricultural award is given to farmers with the highest standards on overall farming practices and improvements.

Poultry farmers in Quebec appear to have a distinct advantage over those in other provinces: one cohesive union, made strong by loyal supporters like Stewart. The fifth generation farmer and third generation union man has a bottom-line approach to farming. "I'm a supporter of the Equitable for Everybody system, where farmers and retailers work together to sell local products at fair prices. I'd just like to see everyone make a decent living."

"Raising chickens is one of the smart options in farming," says Stewart. "You can produce the largest volume of meat in the shortest period of time."

Stewart Humphrey is a proud holder of his UPA (Union des producteurs agricoles) card. Limited to Quebec, it is by far the strongest farm union in Canada with over 40,000 members. Forming strong alliances isn't easy since by their very nature, most farmers are fiercely independent. It can be a challenge to present a collective voice to government and regulatory bodies.

What began as only five acres expanded into a 15-acre-field of peppers. Plans call for a total of 36 acres. For older farmers, the concept of an entire field under one roof with total climate control, is largely unimaginable. The plants are hydroponically fed and can grow up to five metres on their support lines. With the help of an electric picking cart, Scott is able to check the new growth for pests.

NASCAR Peppers

ABBOTSFORD, B.C.

THE GREENHOUSES ARE SPOTLESS. Sterile-looking canopies house long, narrow rows of perfect pepper plants. Each plant is supported by a four-metre line stretching to the ceiling. A maze of thousands of these lines gives an other-world appearance to the lush plants hanging plump with flawless fruit.

As you leave the growing area and enter the packing room, the humidity dissipates and the temperature drops about 16 degrees. Here, graders, packers, and tow-motor drivers tend to the processing lines. Working together, the staff move 7.5 million peppers each year. Just before shipping, the peppers pass through a robotic labeling machine that brands each one with a distinctive sticker: NASCAR Peppers.

The NASCAR Pepper story begins with 57-year-old Don Voth. Don is the son of a Russian Mennonite farmer and builder who, he says, taught him the value of hard work. Don began to

On a given day, a staff of about 25 women will pick approximately 36,300 peppers — enough to supply one family's needs for 279 years!

The Edwards, (left to right): Alexis, Laurel, Melina, Lincoln, Scott. Spraying a chalk solution on the west side of the green houses minimizes the sun's intensity and the likelihood of burning the pepper plants. A quick rub and presto — portrait windows.

develop his business skills as a teenager, selling fruit from his family's farm door-to-door. He learned the meaning of perseverance while working in his dad's construction business. "I realized that being the boss's son wasn't necessarily a good thing when he handed me a really basic implement — called the shovel. My dad always paid me half of what he paid everyone else: half of my money went back home, and the rest I could keep."

Over the next 20 years, Don grew the family construction enterprise into a multi-million dollar operation. Although his passion for building led to phenomenal financial success, he says there was something missing. "I was always building something for other people. I longed to go back to my roots and build

something lasting for my family. I missed farming."

Don says there's something special about moving and owning dirt. When he found an abandoned gravel pit, he knew he had the right location. Four years and many tonnes of dirt later, he'd built 15 acres of greenhouses and 13 acres of cranberry bog in the heart of some of the most sought-after farmland in Canada. And some of the most expensive: in Abbotsford B.C., one acre of land currently sells for around $50,000.

Don's wife, Elma, says that Don's determination and stick-to-itiveness are forces to contend with. "He doesn't know what *no* means," says Elma. "He's tenacious, absolutely tenacious! When someone says something can't be done, that's when he starts getting busy." But the sharp businessman has a somewhat different take. He says the success of Merom Farm comes from focusing on efficiency and surrounding themselves with great people. "We hired the Wayne Gretzky of growers last year and our yield increased by 30 percent. Merom is a family-operated business. Each family member brings their own skills and personality; that's what makes this operation a success."

The family meets on a quarterly basis to review and participate in strategic planning sessions. They've customized a business model that they refer to as REV. The focus of REV is to enhance relationships while effectively maintaining the group of family corporations. Their mission statement provides simple but wise direction. "Get along. Work together. Make money. Provide for needs. Do ministry. Have fun!"

Don's son-in-law Scott is a key player in the operation. He and his wife, Laurel, and their three children live in one of the family residences on the Merom property. Laurel met Scott at bible school and says they were a mismatched couple. "I was the typical good girl. He

Once picked, the peppers tumble gently through a sorter and are sized into four different grades. Farmers are under increasing pressure to produce products of varying shapes, sizes or colours to meet consumer demands.

was the bad boy who was sent to school to shape up. But we slowly got to know each other, and on the very last day of school, he made his move. He came over, poked me in the arm and said, 'We should be together!' That summer, he sold his stereo and bought a bus ticket. The rest is history."

Scott welcomes challenge and change, and appears to be just the right man to manage the pepper facility. He recalls the years of hard work that led up to the launch of the greenhouses. "What a great sense of achievement it was picking that first pepper! Finally, after three-and-a-half years of planning, talking, and working hard, we shipped out the first case. Now we ship 300,000 cases each year. Our Sweet Bells are some of the best you'll find anywhere.

"The pepper cycle is twelve months. We seed in October, plant in mid-December, and harvest from March until November. Our state-of-the-art facility balances the nutrition daily and gives us plants with thick, strong roots and just the right balance of leaves. Right now, the greenhouse is full of rows and rows of brilliant peppers — reds, yellows and oranges."

Merom Farm is positioned as a provider of a superior quality product supported by a strategic brand with world-wide recognition. Scott says the branding strategy reflects the family's personal and business interests. "Some of our family members race — we're huge racing fans. We actually own eight cars ourselves. We are very proud to have the NASCAR branding. Our entire product is shipped to the U.S. and having the NASCAR label makes it more attractive to consumers. It's a brand people can recognize and feel a loyalty to.

"The biggest challenge we face is dealing with increasing exchange rates and high fuel costs. At Merom we are constantly planning and trying to work smarter. But I absolutely love the challenge that Don and his family have offered me here. I see lots of variety and I can use my creative abilities.

"The future for the smaller guys in our industry doesn't look good. Those farmers who started the pepper business are not going to survive unless they go big. It's an unfortunate situation because farmers need to farm — we *need* to be entrepreneurs. You know, you can always tell a farmer when you meet one. They're the good guys!"

Nicholas, age eight, is the farm's appointed animal trainer. By nursing his three little pigs around the track, he'll have them well trained to race for the Oreo cookies waiting for them back in the shed. As one of the newest developments to hit Canadian farms, agri-tourism is an entertaining and profitable way to connect farmers and urbanites.

Agri-Tourism

TRURO, NOVA SCOTIA

EACH SKULL IS CHARGED $10.00 to enter. When you join the lineup of Victims waiting to get in, you are serenaded with the horrific sights and sounds of the 1974 classic movie, *The Texas Chainsaw Massacre*. You proceed through a tunnel so black you can't see the hand in front of your face. But you definitely feel the sweaty dampness of — something — pawing at you! At last, you arrive in an open corn field, only to meet a chorus of blood curdling screams and monsters coming at you with chain saws. Real chain saws. And it's guaranteed to be *the most fun you will ever have in a cornfield!*

You've just experienced one of the newest and fastest growing trends in rural communities today. Agri-Tourism is opening farms to the public for fun and profit all across Canada. With wagon rides, a U-pick operation, a petting zoo or a Halloween corn maze to scare the bejeebers out of you, farmers have begun to move into the world of entertainment and direct marketing.

Jim Lorraine, the son of a retired agriculture politician, is a charming, well-spoken young farmer with a different perspective on farming. "We grow entertainment here at River Breeze Farm. I may market

Early in his farming career, Jim recognized that traditional farming and crops were not enough. His solution? Eighteen acres of "Supersweet" — a gourmet sweet corn. It has rapidly become a perennial favourite. Even priced at $4.00 a dozen, more than twice the going price, his crop sells out every year. Sweet corn accounts for only three percent of the total corn acreage in Canada.

myself as a farmer, but I consider myself to be in agri-business." Jim is a skilled, creative marketer who has the ability to execute and dares to be different. He even does his own voiceovers for radio commercials. Jim laughs when he describes last year's Wild West theme and dressing up as a sheriff. "It's all about having fun — we create a happy experience."

Jim's evolution into direct marketing began quite by accident in the early 1990's. As a beef producer in Truro, Nova Scotia, he was hit hard by poor market prices. He decided to diversify into strawberries. But Jim found it to be hard work for a low rate of return, as most of his product was sold to local stores. On one occasion, he was left with 40 extra flats of strawberries and no one to sell them to. His rapidly ripening product forced him to innovate. Jim loaded his berries onto his flatbed and headed for town. He parked his load on the side of a road and put up a small sign, 'Strawberries for Sale.' "I sold all of those berries out of the back of my truck and was totally surprised. Right then, I decided to go retail and avoid the middlemen. And that's how sixteen acres of strawberries and a U-pick Operation put us on the map!"

His strawberry success was just enough to propel Jim in a whole new direction. He knew he wanted to give customers 'a direct-from-the-farm experience', so he built a roadside market. It is actually a large retail outlet with an in-store bakery and a showcase of high-quality fresh produce.

The Lorraine family (top to bottom): Jim, Tricia, Nicholas, Eryn and Gillian. As part of the farm's agritourism, paying visitors can take the easy wagon-train tour through the 12-acre corn maze. Or, they can wait until the 'special nights', offered just prior to Halloween. On these nights, the effect of scary music, coal-black tunnels and 60 costumed actors — some toting roaring chainsaws — overwhelm the senses. A number of the approximately 3,500 visitors, mostly teens, will hyperventilate, faint, and wet themselves. But they'll return again next year simply to have the bejeebers scared out them.

Jim says one of the key reasons for their success is strategic marketing. River-Breeze is positioned as "An Agri-tainment destination that grows farm-fresh family fun." Most mornings, he can be found at the market office by six o'clock, planning the day's activities, organizing staff, creating sales strategies, and looking after advertising and public relations.

"The farm has come a long way from that first desperate sale of a few strawberries. We now have a 12-acre corn maze, a petting zoo, pig races, and wagon rides. The animals in our petting zoo produce food, clothing, and milk. It's a great way to introduce children to agriculture — an industry that is far removed from the rest of society. They can learn about honey production at our glass-enclosed bee hive or test their navigational skills in the maze. River Breeze is the perfect setting for teaching school children where their food comes from. We are Nova Scotia's living agriculture classroom."

Situated on 600 acres, River Breeze is the winner of the 2005 CNTA Tourism Attraction of the Year Award. The farm has 15 to 80 seasonal employees, raises 300 cattle, and grows sweet corn, pumpkins, sunflowers, feed corn and grain.

Jim, Tricia and their three young children live in the family homestead — a quaint old farmhouse built in the late 1700's. "My father, Ed, was a farmer turned politician and became the Minister of Agriculture for two years. I was the youngest of five kids and the only boy. I took over the farm when I was twenty-two.

"Dad wasn't sold on the entertainment side of farming when I first started, but now he gets it. Can you imagine what he thought when he came by one day and saw me mowing down the corn to make a maze? Farmers grow things, and here I was destroying good corn."

As with most business ventures there was an element of risk, and Jim's major concern was getting enough people to come out to the farm. But show up they did! The River Breeze 'Haunted Halloween Maze' is now Truro's biggest tourist attraction.

Over a three-night haunting, the corn maze attracts 1,200 people each evening. Their catchy slogan, 'You'll pay to get in but you'll PRAY to get out', isn't far from the truth. Jim does not recommend it for children under ten or for the faint of heart. "It's a huge amount of work. Each year we employ sixty actors, six make-up artists, a complex lighting and sound system and lots of staff to keep the lines of people moving. It's a blast!"

Just when you think you know all there is to know about River-Breeze, Jim tells you about the 'Turducken'. "We grow food here for entertainment — like the 'Turducken'. It's a specialty order item that costs $120.00 and customers love it. Our chef de-bones and flattens a fresh whole turkey and layers it in spiced sausage. Then he de-bones and flattens a duck, lays it on top of the turkey with another layer of spiced sausage. But there's more. He does the same to a whole chicken, and then he stitches it all back together so it looks like a turkey. It tastes amazing! And it's a social event that feeds 30 people and takes about 8 hours to cook. Imagine, 15 couples waiting for this wonderful food, something they've never experienced before — a real 'talk about item'. It's food entertainment."

Jim thrives on the freedom the farm offers to create something new and different. His success lies in developing successful marketing and branding strategies. He is intentional about putting a positive spin on everything, and in times when urban folks may feel disconnected from rural life, his farm welcomes you.

As small family farms continue to struggle to survive in a world of specialty markets, low prices and rising costs, more farmers may look at non-traditional methods of farming. It's smart, profitable marketing that entices us to, 'Stop, come play and smell the sunflowers'. And so far, several thousand people have come to do just that.

At 30 kpm and with a 30 metre boom, Colin can spray 800 acres in a good day. Estimated cost of chemicals? Upwards of $24,000!
Decisions around the application of herbicides (weed control), and pesticides (insect control), regarding which chemical,
how much, and when — are some of the most important decisions a cash crop farmer makes in a season.

Succession...

When is it time to sell the family farm?

EDMONTON, ALBERTA

INDUSTRY IS KNOCKING on Colin Jackson's door. The young Alberta farmer is the caretaker of 3,500 acres of prime farmland that has been in his family's safekeeping for 125 years. Hemmed in by urban sprawl and faced with developers bearing enticing offers, Colin has no choice but to explore the possibility of selling the family farm. Depending on your perspective, Colin has either the good fortune or the misfortune to have one of the fastest growing regions in Canada (the Calgary–Edmonton corridor), in his backyard.

About 98 percent of farms in Canada are family-owned and operated. And, according to Statistics Canada, small farms (defined by gross farm revenues of less than $50,000 annually) accounted for more than half of all Canadian farms in 2001.

The average Canadian can anticipate changing their job approximately every nine years and moving house and home several times during their lifetime. Such lifestyle changes offer the opportunity to reinvent oneself in a world where moving on is equated with moving forward. For farmers, these changes

*Richard and Rita Jackson's retirement home is only three kilometres from the old homestead, so return visits are easy.
"Those were good years," says Rita. "And when I'm here now, I simply feel thankful. Thankful for how well
the farm is managed by the boys, the grain bins full, and the young people riding horses — just like we used to."*

are much more complex. A farmer's lifestyle, family, business, and history, are tightly interwoven. The farm is the family's heart and identity. Farmers inherit the sweat equity of grandparents, parents and siblings. So the question of Colin's selling or not selling the family farm is emotionally charged, and goes right to the core of a family heritage.

In 1993, tragedy struck. Colin had graduated from film school and was beginning a career in the world of film making. Suddenly, his father Dick was seriously injured in a shop accident: he could no longer manage the farm. Along with the anxiety around Dick's health, the Jackson family was confronted with the sudden loss of the farm's steward. They rallied, reviewed their options, and decided to implement a succession plan. Colin was forced to make a career decision. Here was the opportunity to own his own business and at the same time, fulfill a family obligation. At 26 years old, he left his job to farm full-time with his brother Tom.

The partnership proved to be prosperous. In addition to working the family land, the brothers rent another 1,500 acres to grow wheat, winter wheat, canola and saba beans. Their fields yield about 140,000 bushels and generate annual gross sales of $1.3 million.

Alberta is Canada's number one wheat producer; wheat and canola are the two leading crops. Farmers like the Jacksons are feeling the pressure to go big or go home. Colin's morning typically starts with phone calls and office chores. Once he heads off to the field, he is often there until sundown. "We are competing on an international level and it gets more difficult each year. I don't like what farming is turning into. In the future, will we all grow 20,000 acres and become corporate operations? Seventy percent of our grain is spoken for by June, and completely sold by the end of the year. When shopping our product, we aren't looking for the highest dollar, but just enough to make a profit and continue on next year. We look for assured profit, not maximum profit!"

According to Colin, buyers have the ability to custom order. "Our focus is on growing specialty crops for a specific buyer. Grain is grown and sold all over the world. The Japanese are

Colin Jackson loves technology. Operating his new $300,000 sprayer is more akin to flying an airplane than driving a tractor. For the most part, he focuses on monitoring two video screens, while the automatic, razor-sharp GPS steering keeps him within 10 centimetres of his intended path. "Environmentally, this is a better way," states Colin. "I can spray more precisely using fewer chemicals. I'll use any technology that helps me be a better steward of the land."

buying our canola." Colin's customers represent ever-changing consumer preferences. Buyers may specify a particular protein or grade. They may request, for example, a hard, white wheat that yields a paler kernel, producing a healthier, but whiter-looking whole-wheat bread.

The Jackson family farm appears to be prominent on the oil industry and developers' radar screens. Their desire to discuss options has certainly given Colin food for thought. But how do you place a value on the serenity of a perfect stand of golden wheat, undamaged by the wind? Or on the stunning vibrancy of a yellow canola field in full bloom?

"I guess I could use this opportunity to try something different, possibly buy a smaller farm somewhere else. But I live in a great, tight-knit community — the kind of place where you don't have to lock your door. And when things aren't going well, I have family and friends here who understand what I'm going through. We're in this together. There is no reset button in farming. I can't just get up and walk away — I have to carry on. And I love the fact that my work is in my backyard. When I step outside my door, *I know exactly who I am, and what I do.*"

When grandson Nikolas, age three, is ready for a mid-afternoon snack, he knows Nick expects everything to be shared equally.

Protecting Our Natural Gardens

JACQUES LAFORGE, A DAIRY FARMER in Grand Falls, N.B., has a message for Canadians.

"Canada is very fortunate. We have one of the safest, healthiest, highest-quality food sources in the world." And he should know. This is a farmer whose voice is heard all around the world.

As president of The Dairy Farmers of Canada, Jacques has direct input into policy-making decisions in the world dairy industry. While other countries around the world are just beginning to look at food safety issues, Jacques is confident of our situation at home. "Consumers should understand that farmland is their land. We need to work hard to protect our national gardens. We have a high-safety food program in Canada that is completely traceable. If a situation were to happen, we could react immediately.

"This is not the case in all countries. In 2004, dozens of babies in China died, and thousands more were seriously infected, due to malnutrition related to limited food safety regulations. Babies being fed formula were starving to death. It was later discovered that, in an attempt to save money, processors had altered the milk, taking all the nutrition out of the formula. The government investigated and initiated regulations but only after many infant fatalities. This could never happen in Canada. Because of our high standards, the world is now looking to us as an example," Jacques says.

On any given day you might find Jacques either on his tractor cutting hay or delivering a presentation to a delegation in Mexico. His personal calendar reads like an international travel log: *Tuesday — Toronto; Wednesday — Ottawa; Friday — Geneva.* Dividing his time between running his own dairy operation and

Patsy and Jacques Laforge and some of their 1,000 acres of undulating fields by the St. John River.

The Laforge grandchildren, like all Canadians, are fortunate in having access to an abundant supply of some of the world's best-quality milk.

Jacques and wife Patsy along with their son and daughter-in-law, Rock and Melany, own 1,000 acres along the scenic St. John River Valley. Their dairy farm, oddly enough, is located in the middle of their potato fields. Along with milking eighty-five Holsteins, the Laforge family grows barley, hay, corn and alfalfa.

Patsy works full-time off of the farm, which Jacques describes as the family's insurance policy. "Throughout the years, Patsy's job has allowed us more freedom and given us the ability to take more risks in the business," he says. "We have also recently expanded into the trucking business. We already had the equipment for the dairy business and wanted to use it more efficiently so we developed a trucking market. We bought a transport truck and now do our own hauling. We sell hay and straw locally, and also ship to a market we've developed in the U.S.

running to catch an international flight is old hat for this experienced Ag politician. "You have to be half crazy to take on this position. I can't recall the last time I've been home for a whole week," he says.

"I became involved in politics because I wanted to have an influence on dairy farming." Jacques speaks with the same confidence and ease whether he's helping a local farmer in trouble, or discussing the complexity of international agriculture at a WTO (World Trade Organization) hearing. He says the most critical function in his role as president is communication: ensuring that consumers and the general public understand the importance of dairy farming as a commodity. His knowledgeable, down-to-earth approach, makes it obvious why this soft spoken, unassuming farmer is now in his third term as president.

"One of the major challenges we face now is escalating costs. Land prices have gone up. An acre around here now costs $4,000. And we're not getting good prices for our products. To keep up, farmers are forced to produce greater volumes and get bigger. And that means taking on more financial pressure," he explains.

"Over the next 20 years, I see increased economic pressures for farmers. They'll be forced to become even bigger operators. I think we'll face more environmental and animal welfare issues. The good news is that technology has changed farming in many positive ways. We have more mechanization, allowing us to farm more acres with less human labour. Here on our farm, we now have a modern milking parlor. If we compared production to the 1980's, our cows are producing three times as much milk. The bottom line is that technology means higher production, less labour, less time."

Jacques Laforge considers it a welcome break to occasionally get back in his tractor. As president of the Dairy Farmers of Canada, he's on the road more than half the year, meeting with other farmers, talking to politicians in Ottawa and travelling around the globe. His mandate as an Ag-politician is two fold: working with the World Trade Organization to protect Canadian dairy farmers through their supply management system and, ensuring those same farmers are not being undermined by Canada importing substitute — and subsidized — dairy products.

As a third-generation New Brunswick farmer, Jacques has many fond memories of being raised on a farm. "There's nothing more gratifying for a child than getting on the tractor with your dad. When I was really young, I remember picking potatoes out in the field. It was a community affair. All the adults and kids worked together picking the potatoes by hand, putting them into the baskets that we dragged through the field. It was great!"

Jacques has a different take on what can be a farmer's most challenging obstacle — uncontrollable weather — and articulates a simple but profound truth. "I love operating an independent business that is linked to the natural environment, and success that is linked to your ability to interact with the climate. But one thing is true: If there's money to be made in farm production, the farmer will be the last to make it. And if there is money to be lost, the farmer is the first to lose it!"

Like many farm operations today, the Keefers rely on more than one income source.
Donna Keefer has been selling hanging flower baskets for more than 20 years, supplying about a third of the family's net income.

Cranberry Fields Forever

RICHMOND, B.C.

EVERY SEPTEMBER, Dan Keefer loves to watch his cranberry fields transform themselves from leafy vines sprouting pink flowers, to acres of dark-red berries. When the fruit is at its peak, he will flood the fields and create a mass of floating berries. Can you imagine a barefoot walk on this crimson carpet? It's one that Dan has taken many times. He says there's nothing like a fall harvest stroll on *a bright red sea of wet marbles*.

Cranberries grow on trailing vines in low-lying beds and require acid peat soil, an extended growing season, and a good fresh water supply. At harvest time, these beds, also known as bogs, are flooded with about 15 centimetres of water and the fruit is beaten off the vine using a special harvester. The floating berries are then corralled to one corner of the bog before being pumped or elevated into a truck.

Cranberries were first used by Native Americans who discovered the berry's versatility as a fabric dye and a food, which they believed had healing properties. The versatile tangy tart fruit was first cultivated commercially in 1816. Research in recent years has analyzed its powerful natural compounds and validated the power of its nutrients — notably its antioxidants. Cranberries have moved beyond a Thanksgiving tradition and are now found in over 500 different food products.

The Keefer family, (left to right): Jake, son Grant, daughter-in-law Justine, Donna, Dan and daughter Teresa.

Dan and his son Grant inspect one of their spring cranberry fields for winter frost damage and insect stress.
Although hemmed in all sides by city development, their land is protected from development, at least for the moment, by the Agricultural Land Reserves.
But land is severely limited and with prices pushing $30,000 an acre, it's questionable whether the next generation can afford to carry on.

Dan's 75-acre farm is located in Richmond, B.C. He and his wife Donna employ 10 seasonal employees who help grow cranberries, run a greenhouse flower operation, and package soil for garden centres. Their summer open house sells out each year as hundreds of customers line up to purchase more than 3,500 hanging flower baskets. "For the six weeks we're open to the public, I work 12 hour days organizing workers, waiting on customers, working with the plants—and in between all of that, I try to find time to cook the family meals. The soil-packaging part of our business has grown like crazy. It's not glamorous, but it's profitable," says Donna.

Donna inherited her green thumb and her passion for plants from her father. "My dad had greenhouses when I was a girl and he taught me how to care for plants. I've always loved the creative process of designing flower baskets and trying new things. Plants have different habits and you have to know what should be grown together and what will thrive."

Donna notes an interesting difference in how consumers perceive the value of decorative flowers versus food. "It's strange: people will happily pay $50.00 for a cedar basket of flowers and not think about it. But then they complain about paying $5.00 for a small box of peas or tomatoes. If only they knew how much hard work went into growing those peas."

Dan speaks with a calmness and quiet confidence that comes from a long journey of overcoming personal and profes-sional obstacles. "Mom and dad had a tough go of it. When they started out, they didn't have farming experience, machinery, money or family support. The farm was nothing but 75 acres of bush. I was the youngest of three kids and when my brother and sister left, I stayed on even though there was very little hope of making it. I think everyone expected the farm to fail. I guess I stayed because I thought a bad farm was better than no farm at all.

"My mom was a saviour. She taught me not to measure success by other people's standards, but by looking back and seeing how far I had come. If I got down, she'd say, 'Remember what it was like just five years ago and how much we've gone through as a family.'"

Dan says that the topic of retirement is one he and Donna haven't gotten around to. "I guess we will have to deal with the retiring issue one day. One option would be to sell our land, but it's strange — it would be like selling your soul."

The Keefer farmhouse accommodates Phyllis, Dan's 86 year-old mother. Phyllis, who lives in the granny flat, says full credit goes to Dan for rescuing a dying farm to support his family. "Dan made this farm what it is. In the early days he had everything against him and didn't have any help. But he never gave up. The farm was built by his strength, will, his great per-severance, and his desire to not be beholden to anyone."

At the top of the 36-metre-high Elliott elevator are two distributor rooms. Grain is elevated from the grain pit below into this octopus arrangement of pipes.
By means of several control wheels located on the main floor, grains are directed into one of the 36 bins. "The worst thing you can do," says Larrie,
"Is to dump two different grains in the same bin. We had a few goofs in the beginning but we've got it figured out now."

Elliott's Elevator

RESTON, MANITOBA

I F CURIOSITY SHOULD COMPEL YOU to step inside the belly of a giant grain elevator, consider visiting Inglis, Manitoba. For a donation of a few dollars, you can reconnect with prairie life as you tour these original prairie icons. The Inglis elevators provide exhibits, historical photos, picnic areas, and even a souvenir shop.

Stepping inside the Elliott Elevator is a *little* different. As you come into the entrance, wheat dust hangs in the air like liquid-dough. The main door is cavernous, big enough to accommodate a full-length, tractor-trailer truck. You get a sense of the structure's unique function when you walk across an extensive, steel floor grate. Here, trucks dump their grain into the grain pit below. From the pit, a vertical elevator transports the grain to the distributor room at the very top of the structure. The distributor, an octopus arrangement of pipes, allocates the grain to one of 36 separate storage bins. But how do people get to the top of this imposing building?

Back on the main level, a hallway gives access to a one-man

Kelsey Elliott reaches for a grain sample. As each truck load of grain is dumped into the elevator, it is subject to a critical moisture test. Wheat, for example, must be below 14.5 percent moisture. If too wet, it will begin to heat. And if not noticed soon enough, the wheat will turn moldy, becoming worthless even for livestock feed.

The Elliotts, (left to right): Kelsey, Evan, Chris, Taylor, Laura Ellen and Larrie. Once the quintessential prairie icon, wooden grain elevators dominated the skyline. When the Reston elevator was retired by its corporate owners, Larrie Elliott and his brother managed a successful ownership bid. It now houses about half of the 300,000 bushels of the wheat, canola, oats, barley and peas produced annually on the Elliott farms.

lift. The wooden, coffin-like box offers a snug, self-propelled ride that is not for the faint-of-heart. The lift is completely open on both sides. To move up, you pull down hard, hand over hand, on a thick rope. Your speed is controlled by how quickly you pull. But if going up is manageable, coming down is tricky. You reverse the rope direction and subtly apply foot pressure to a floor brake. It's a great workout for the solar plexus as you

control — barely — your plummeting body down a very *quick* 36 metres!

The elevator was the original commercial grain-storage facility for the town of Reston. When it came up for private sale in 1996, Larrie Elliott says it was a deal he and his brother couldn't resist. The elevator would provide ideal storage for the wheat, canola, oats and peas grown on their 7,000 acres.

In the early 1900's, farmers brought their grain to town by horse and wagon. Older elevators like the Elliott's were originally built along railroad tracks so grain could be collected and shipped by railcar. In the 1980's, grain companies began constructing high-capacity terminals in centralized locations, making the out-dated, small-town elevators obsolete.

Larrie says farmland in the Reston area is made up of rolling hills, pot holes, slews, and lots of poplar trees. "It is good land for growing grain but we've had more than our fair share of weather challenges. The last few years we have seen lots of moisture. At one point last year we had more than one-quarter of our land underwater. And this year, in the month of June alone, it rained about two feet."

The Elliott operation produces and hauls 300,000 bushels of grain each year. For Larrie, part of the joy of farming comes from working large, powerful farm equipment of which he has no shortage. His farm boasts: one sprayer, two seeders, three combines, three trucks, four tractors and two dirt bikes. "As a kid, I remember sitting on a combine and doing a man's job. That felt great! I love working the equipment and the feeling of power. It's like driving a fast car."

Larrie met Laura Ellen for the first time at a dance. She was playing drums in her family's band, *The Roach Rhythm Ramblers*. Now, married for 24 years with four teenage boys, they are a family who work, play, and fly together. Larrie and two of the boys are licensed to fly the family's four-seat Cessna.

Laura Ellen works part-time as a kindergarten teacher, and knows that her children understand the meaning of hard work.

"It was easier to raise our children on the farm — I don't think I could have contained four boys in town. Living here, I don't have to worry about where they are and what they're up to.

"At seeding and harvest time, the kids help with stone picking, combining, swathing, rolling, harrowing, planting and seeding," says Laura Ellen. Farm kids are hard-working kids. During harvest for example, my oldest son Evan will work about 100 hours a week. The school and the teachers accommodate farm kids like him in the busy seasons by giving them less homework."

In June, during seeding month, Larrie also puts in long hours, often working until well after dark. Although he feels positive about his family's future, Larrie says farmers today have a tough go of it. "I know a number of farmers who have been forced out or who decided to get out because it wasn't worth it. The trouble with small communities like ours is that everybody knows everybody. It's difficult when you lose the farm and owe people in your own town. Those families really suffer. Every farmer who's survived the last 10 years is an optimist. We have no choice — we have to be. Right now, I get less money for the wheat that goes in a loaf of bread than the packaging company gets for the package."

Larrie begins each day at 7:00 a.m. with a coffee at the local Dennis County Café, where he meets his farming buddies to swap stories and shoot the breeze. When asked what advice he has for other farmers, the hard-working farmer gives an unexpected answer: "Take the time to go flying! *Even* when you should be working."

To survive in today's farming market with its high input costs and slim returns, farmers have been forced to become creative. David Coburn has both diversified and specialized, learning how to turn manure into natural gold. After 42 days in this 75-metre-long composting pit, the resultant compost is ready for bagging.

Natural Gold

FREDERICTON, NEW BRUNSWICK

NOT ALL MANURE is created equal. Depending on the animal and its diet, the quality of the finished product varies considerably. One of the lesser-known, but prime-grade manures is bat guano. Cows and horses are probably the most common sources, and can be relied upon for good quantity. Rabbits deliver an excellent product, and more neatly packaged, though quantity is an issue. But according to New Brunswick farmer David Coburn, the perfect source is Shaver Hens.

When David could no longer make a profit from the table-apple market, he created a niche business manufacturing composting. His Natural Gold product is chicken manure mixed with his farm's organic waste, straw, and sawdust. The Coburn farm also produces apples, cider and eggs from the poultry operation.

What do apples, composting, and eggs have in common? For David, they represent a solution.

"It's in the tough times that we find out what we're made of. My family has been in the apple business for generations, but in 1996, when we lost our $110,000 bumper crop to hail, that was the last straw. Ten years ago there were eight to 10 apple growers in our community: now there's only one full-time producer. Economic realities force diversity upon us. The farm is in a constant evolution, and I'm just along for the ride. People thought I was nuts when I first bought my composter in 1993. Now, I'm right on the money!"

D. H. Lawrence once said, "The fairest thing in nature, a flower, still has its roots in earth and *manure*." Six-hundred

David and Karen Coburn, and their nostalgic connection with yesteryear — a horse-drawn, 1910 "Hardie" sprayer.

The mixture is simple: roughly equal parts of chicken manure, straw, and sawdust — with old apples and a few cracked eggs thrown in for a garnish. At $5.00 a bag, gardeners think it's a bargain. On the other hand, farmers can't believe anyone would pay for a pile of manure. Son Glen is simply happy to collect $7.00 an hour working for his dad.

tonnes of the Shaver's contribution to the *fairest thing* are sold commercially or spread on the Coburn farm.

Innovative and trailblazing are apt descriptors for David's farming style. Along with a number of other agricultural achievements, David and his wife Karen were awarded the 1995 Outstanding Young Farmers prize for Atlantic Canada. In 1989, David was the first in Canada to install a computerized egg barn. This sophisticated system ensures the animals receive proper food and water rations. Sensors scan the barn 288 times a day, continually monitoring humidity, ammonia levels, and the temperature, inside and out.

In peak production, the chickens provide 24,000 eggs a day. David relies on one full-time barn manager to oversee daily egg activity. The eggs are shipped to a grading station three times a week, and then distributed throughout the province. According to David, a typical good-looking Shaver waddles in at around four pounds and is bright white with a red comb.

David says his farm hasn't used antibiotics for over 23 years. "I'm proud to say I raise birds without the use of antibiotics, and that our hens receive a high quality feed with no animal by-products." His flock of 25,000 birds have a healthy—but costly—appetite. In 1995, David built a feed mill and now mixes 925 tonnes a year. His costs are less than they would be buying the feed elsewhere. It's a win-win situation: David supports other farmers by buying local grain and his lucky birds benefit from a recipe of all-natural feed.

David and his spouse, Karen, have three teenagers, Jennifer, Glen, and Tyler, who all help out on the farm. David is a sixth-generation farmer who knew at the age of 14 that he'd rather get dirty than work in an office. "Farming's been in my heart since I was a kid. My first memory is from when I was around four years old. Every Friday our family would kill chick-

Always the innovator, in 1989 David constructed the first computerized egg production facility in all of Canada. Every five minutes, the computer scans the barn for temperature, humidity, ammonia levels, feed and water consumption, and then adjusts the inputs accordingly. Twenty-five-thousand chickens lay almost the same number of eggs each day. Daughter Jennifer checks for cracks before the eggs move on to an automatic packing machine.

ens and they would let me sit and pluck with the grownups, like everyone else. My grandmother taught me how to gather eggs by hand. When I got a little older, I pestered the flock manager until he let me work with him."

The Coburn farm has been in David's family for 200 years. It's a responsibility that David clearly loves. "To know where you are going, you have to know where you came from. It's good to know your roots. At times, I have the feeling that my ancestors are still here on the land. And for now, I'm just the caretaker here."

The Gartrells: Dave and Pat. Pat was only 20 when she fashioned her own wedding dress. "I told Dave I'd marry him
as long as I didn't have to do the farming as well as raise a family. The kids are so grateful now that they always had their parents at home."

Local Means Tastier

SUMMERLAND, BRITISH COLUMBIA

"A SOD-ROOFED CABIN with dirt floor was home; birch twigs lashed together a broom; a saucer of grease with string wick the evening light." This excerpt from historian Mary Gartrell Orr's journal, paints a vivid picture of her log-cabin life in 1887. Mary, the first white woman settler on the west side of the Okanagan Lake, in B.C., helped establish the first commercial orchard in that area. She was also the great grandmother of apple-grower Dave Gartrell.

A true pioneer, strong and self-sufficient, Mary described the Okanagan area in her journal as *The land of plenty.* Her love

for the land has been passed down from one generation to the next. "My dad felt that land was the greatest asset you could have," says Dave. "He inherited that passion, that love from his parents. One of his favourite sayings was, 'They're not making anymore land you know!'"

"I'm the fourth generation of apple growers and I've always enjoyed being in the orchard. When I was a kid the trees were huge—20 feet high. When the pickers were finished for the day, my brother Fred and I would climb to the very top and get the apples they'd missed. We had great apple fights."

More than 120 years ago, Dave's great grandfather used this fruit wagon to haul peaches and apples to the mining camps at Fairview — a three day trip. "Seeing this wagon connects me immediately to my past," says Dave. "Those farmers worked unbelievably hard."

"I guess I inherited my dad's pioneer instinct, his way of being. The transition to go into farming was interesting, but not exactly in the way you might expect. I was young, living at home, and one night I came down to the kitchen. My dad was sitting at the kitchen table. He handed me a lease agreement for the farm—written on a table napkin. And that was it! Now, I farm with my brother Fred, the eternal optimist. And as I'm a pessimist, it works out well because he's always there to pick me up!"

Dave, his wife Pat, and brother Fred, own and operate a 35-acre apple orchard in the heart of Summerland. Summerland lies in the interior of British Columbia approximately 425 kilometres from Vancouver. Located within the Thompson–Okanagan Plateau ecoregion, it has one of warmest and driest climates in Canada. Its agricultural industry is focused on tree fruits — apples, peaches, cherries — and grape production.

The art of growing apples has changed dramatically in the last 30 years. Dave, like most growers, uses the Super Spindle System. It's a high-density planting and growing technique with 2,500 trees to the acre. The trees are much smaller than in the past, growing only two and a half metres high, with a three centimetre diameter trunk, and branches half that thickness. Each tree yields about 40 apples.

Harvest on the Gartrell farm starts in September and lasts about eight weeks. During picking season, Dave works 12 hours a day, seven days a week. The apples are picked into 360-kg apple bins, and then trucked off to a warehouse. There, they will be graded and then sold over the next six months.

His crew of 10 to 15 pickers, made up mostly of East Indian women, arrives at six o'clock each morning. "The women are excellent workers. We have great respect for them — they have a wonderful work ethic and we look forward to them coming back each year. I would like to pay them more, but our biggest challenge is competing with the cheap fruit that comes in from Washington. We have a tough time as it is. Although minimum wage is $8.50 hourly in B.C., we pay $11.00 an hour. We also work in a bonus system.

"Ambrosia apples are our latest success story. It's a cross between golden and red delicious. In the apple business, the risk in trying something new is quite high so it's tough to decide which varieties to go with. We have so much competition from U.S. growers; Ambrosia is the only one we can make money on. Washington dominates in all of the other varieties."

Dave's brother Fred suggests that consumers could help farmers by choosing to buy more local product instead of imported. "Four years ago our wholesale price for Royal Gala was 30 cents a pound. Now we get 18 cents and we have no control. B.C. grows four million 40-pound boxes of apples. Washington State grows 105 million. They can produce cheaper because of lower land costs and using labour from Mexico. They get their water from B.C., and they also receive agriculture subsidies that farmers here do not get."

Dave agrees emphatically: consumers have the necessary power to change the situation. "We can produce 95 percent of Canada's apples. If Canadians chose to buy only local apples, we wouldn't have an issue. People should try to understand the value of eating local. The U.S. is shipping apples here that have been sprayed with pesticides that have been banned in Canada."

Dave and Pat started dating in grade 10 and their friendship quickly blossomed into love. They've been married for 33 years and have five grown daughters and four granddaughters. When the children were old enough, Pat went back to teaching. All five girls have since graduated from the high school she taught in. "Dave farmed apples, I farmed children! It's a great lifestyle here. Our home has a heart and you feel it when you come in. Dave is a wonderful dad — he's a caring, compassionate guy. What you see is real."

Dave says money isn't the most important thing. "If our kids carry on in farming and make it their way of life, the rewards will be there. We are a small operation and try to grow the best fruit we can. It's about quality not quantity. And beyond food safety and issues of control, the ultimate reason for choosing our apples is taste. Local means tastier!"

Dave does speak the truth. One bite of a Gartrell apple is an explosion of delight. The apple skin breaks crisply and releases the fruit's sweet, satisfying flavour. Here, local means one of the finest apples anywhere.

Disaster Album

EDSON, ALBERTA

WINSTON CHURCHILL SAID, "If you're going through hell, keep going." He was referring to those times in life when what *can* go wrong, *will* go wrong. For Jill and Harv Depee, life on the farm has on more than one occasion gone wrong. Jill laughs as she shares what she calls, *The family's personal Disaster Album*. The album is a visual record of the many unexpected challenges and tragedies her family has endured. It is also an uplifting story of a husband and wife's journey of overcoming adversity and working together as a team.

Jill met Harv when he was already well into his career as a farmer, and she was only 15. "I was walking down the street with my best friend Lori when she pointed at a young man in a black Trans Am and said, 'That guy is so cute!' I marched right out into the street and waved him down. It didn't click for Lori, but Harv and I sure made a connection! A few weeks later when he came to pick me up for our first date, he was holding his arm out like he was hurt. He told me that he'd fallen out of a rutabaga tree. I believed him. He kept that joke going for two months. It seemed like everyone in town except me knew that rutabagas grow on the ground!"

The first few photos in Jill's album show fields of damaged crops littered with big white chunks of ice. "We were only married for one year when we had a 20-minute hail storm that decimated everything. The hail was the size of golf balls, and totally wiped us out."

The second disaster took place just two years later. "It seemed that we were just recovering from the hail when we had *The Fire*. Our equipment, vehicles, crops and buildings were all lost. The fire destroyed 120 tonnes of our rutabagas and potatoes, and we suffered $400,000 in damage. That was really traumatic."

Each year in Canada, there are about 800 fire-related deaths and many of these happen on farms. The financial loss on a farm can be much higher than in an urban residence. Farms have several buildings and are more vulnerable because of their relative remoteness from fire stations.

A few years later, Mother Nature once again took its toll, this time in the form of huge snow drifts and howling blizzards. "These photos are of *The Big Storm*. That year we had such an unusual amount of heavy snow that it collapsed our storage facility and caused major damage to the buildings."

Irrigating 50 acres of rutabagas takes several days of just slugging it out. "These pipes weigh 18 kg," says Derek, "And by the end of the day, you're wiped. I certainly don't need to go the gym to keep in shape."

As if hail, fire, and blizzards weren't enough, the variability of crop prices, insects and fauna have also been challenges to contend with. The deer and elk in Edson, Alberta are more than plentiful, and love rutabagas. "Wildlife has been a huge problem here," says Jill. "One year the elk destroyed 40 percent of our rutabaga crop. We had to put up an eight-foot fence around 160 acres of our fields. Then the following year we lost 80 percent of our potato crop to blight."

But Jill's sense of humour and optimistic spirit is contagious. As she reaches the last few pages of the album, she laughs again. "You know, there's no more room left in our book. Maybe that means we are done with disasters!"

The Depee family farms 50 acres of rutabagas and 80 acres of potatoes. "Harv and I are a team. Along with looking after the kids, I help out wherever I can. Whether it's taking care of the financial books, feeding the hired hands, or washing rutabagas, we work the farm together."

Jill credits having a strong mother as her inspiration. "Our family faced some really difficult times when I was growing up, but my mom always pulled herself and the family up. She did whatever she had to do — she took on the hard jobs, took care of us kids, and made sure there was food on the table. My mom always hung in there, and at the same time managed to upgrade her education and make a better life for us. She taught me a lot, and I'm very proud of her.

"The most important job here for me is to keep Harv grounded," reflects Jill. "Farming is super stressful and he puts so much into his crops. We both work to keep the family going. It's a great way of life and a good place for kids. When our first boy, Derek, was just starting to walk, he loved to be with his dad. When Harv came home for a coffee break, Derek would block the door so his dad either had to take him out to the field or stay home and play."

If you ask Harv what keeps him going, he doesn't miss a beat: "It's Jill. I couldn't do this without her — she keeps me focused on the bright side." Harv is a self-made man who quit school at the age of 15 to grow a few acres of potatoes with his brother. "I worked hard on the farm, and just felt that school wasn't for me. But it was a great feeling when I bought out my dad at 19 years old. It felt like a real accomplishment."

Rutabaga comes from the Swedish word *rotabagge*. The bulbous root vegetable is a cross between the turnip and kale, and is often served mashed or in stews. Rutabagas grow from a small seed to a little plant within ten days. Its dark green leaves grow up to 90 centimetres long. Plants produce hearty purple rutabagas about 13 centimetres in diameter.

The Depees, (left to right): Harv, Shantel, Shaylyn, Jill, Derek, Shelby, and Dylan. The Depee family take seriously the importance of playing together. "It's something we've always done", says Jill, "Whether its monopoly or hockey. Although it's not my favourite, they even talk me into playing touch football now and then."

Harv's fields yield 25 tonnes of vegetables per acre. The fields are irrigated by a hand-move system — a labour intensive process that requires manually moving pipelines. His potatoes are sold locally while the rutabagas go to a co-op in Edmonton for distribution throughout Western Canada.

"What makes us different is that we've eliminated a few of the middle men," says Harv. "I'm an independent operator and I have a small processing plant that can wash, grade, and pack our vegetables. I can deliver the product myself to various markets. Our vegetables aren't sold on contract: we grow hoping to sell whatever we produce.

Eleven-year-old Dylan inspects rutabagas in his dad's field. For the first few weeks after planting, rutabagas need daily checking, particularly for flea beetles. Just a few hours of a warm wind can infest the crop with these insidious pests, which can in turn destroy an entire field within two days.

"It would probably surprise most people to know how hard I work to produce a crop. They see a one-kilogram rutabaga and have no idea how much time, worry, and sweat I've put into it! I think, overall, farmers will have a tougher time in the future. With the cost of land, fuel, equipment and labour rising, it's just going to get harder. But right now, I am my own boss — making decisions, and creating my own future."

Harv and Jill are tenacious and have a perspective on life that many farmers share. Although they have had more than their fair share of life's challenges, this young couple remains grounded and full of optimism. "I think the definition of success is always being honest with people, and doing the best you can. If you have that, then I think you're successful.

"We love our life. There is so much beauty here—the river wraps around our property. It's everything you'd want in a perfect campground and it's all in our backyard. This is our livelihood and a wonderful place to live. We are blessed."

Wild blueberries are one of only three crops indigenous to North America (cranberries and Concord grapes being the other two).
Clear the trees and underbrush and if the soil's acidity is right, the reward is a patch of wild blueberries. Although more labour-intensive
to grow and harvest, wild berries are more richly flavoured and contain more antioxidants than their high-bush, cultivated cousins.

Buzzin' the Berries

CAN YOU IMAGINE strolling through luscious wild blueberry bushes with millions of honeybees buzzing all around you? Visit Russ Hawkins at his blueberry farm in spring and you'll discover fields that are 'just a buzzin!' The distinctive sound is made by 10 million bees stroking their four tiny wings 1,400 times per minute. Russ rents 200 beehives from a local beekeeper at a cost of $80 per hive.

Each hive is home to 50,000 bees who will work for Russ for a three-week period. The bees help bring fruit to flower by pollinating the delicate bell-shaped blossoms of the berry bushes. Each hive is led by a queen bee who lays her eggs in spring and who has thousands of loyal workers to tend to them. The direct value of honeybee pollination to agriculture in Canada is about $1 billion dollars annually. Eighty percent of the pollination of all fruits, vegetables, and seed crops is accomplished by honeybees.

The Hawkins annually produce 12,000 bottles, in four different varieties, of their own blueberry wine.
It has a fresh, clean flavour, and leaves a tantalizingly distinct aftertaste of blueberries.

Donna and Russell Hawkins enjoying the quintessential blueberry dessert experience:
blueberry wine, cake, fruit, and centerpiece. And perhaps "Blueberry Hill" playing softly on the CD.

North America is the largest producer of blueberries and accounts for about 90 percent of the world's crops. The Hawkins farm has the capacity to produce one-half million pounds of the fruit. Blueberry plants like acidic soil, and according to Russ, they'll grow in sand, gravel — or wherever you clear land. "Blueberries are a two-year crop so you have face mother nature not once, but two years running. As well as using the services of a beekeeper, we work hard to keep our native pollinators. Other than organic farmers, we use fewer chemicals than any other agriculture product on the market. Blueberries are a clean product.

"To prune the plants we do a spring burn. It's actually kind of fun. We use furnace oil in burners that shoot out 15-foot flames to burn plants right down to the ground. There's lots of smoke and flame — it is an amazing sight to see! The following year, a new plant comes up that will bear fruit that summer. We generally keep 200 acres in crop and 200 acres in sprout.

"Up until the 1990's, we harvested the berries manually. In

those days workers used blueberry rakes and buckets — it was hard work. Then we went mechanical and bought three tractor harvesters that require only a handful of people to operate. We went from 75 people with rakes down to six people who operate equipment."

Wild blueberries have grown naturally for thousands of years. They are one of the few berries native to North America and are Charlotte County's (Russ' hometown) dominant fruit crop. Along with the rest of the county, Russ faces a major challenge in the weather. Snow protects the plants during the winter months. With the last few years in New Brunswick being open, (little snow ground cover) and cold, producing a great crop has been more difficult.

But they are a great crop. Ever wonder why blueberries are blue? Their colour comes from anthocyanin (a water-soluble pigment), making them one of the few "true blue" foods. *Wild Blueberries primarily spread by rhizomes, or underground runners.* At harvest time the bushes are 20 centimetres high with small teardrop leaves bearing sweet strings of powder blue berries. The sweet and tangy little berries not only have a big taste, but are deemed one of the world's new super foods. Rich in natural blue antioxidants, wild blueberries have been shown by research to provide protection against major diseases, and the effects of aging. The magic of blueberries has been known for some time: they were given to pilots during WW II to promote good night vision.

Canadians aren't the only lovers of blueberries: bears have a reputation for being the real connoisseurs. Bears in the wild can be discriminating, and will select only the most succulent, juicy blueberries of the season. They will travel up to 25 kilometres a day on an empty stomach to gorge on a blueberry patch.

Russ and Donna Hawkins are the epitome of 'down home' hospitality. Donna describes their becoming reacquainted in their home town, years after they'd first met. "When I first saw Russ again, I thought, man, he is handsome! I was attracted to his quietness, and just sensed that he was a good person." Russ is a third-generation blueberry grower. He and Donna work as a team to run their 400-acre blueberry operation and winery.

They have an on-site processing plant to wash, freeze, and ship domestically, to the U.S., and to the European market. The bulk of the berries end up baked into pies and muffins. Their blueberry wine is sold at restaurants on the east coast, and also directly out of the winery. 'Hian Dri' (named for its higher 14 percent alcohol content) and 'Bonny River Blue' wines have developed a great local following. Blueberries produce a sweet juice with a pleasant bouquet. And it has a very distinct after taste, one unlike any grape wine.

Wild blueberries are fresh-frozen at harvest and make a perfect healthy snack that is packed full of antioxidant goodness. Just a one cup serving contains about 18 mg of vitamin 'C' and high amounts of dietary fibre. One of Donna's favourite recipes is 'Wild Blueberry Ice'. It's simple, fast, and easy. "Take 2 cups of frozen wild blueberries, one-half cup of sugar, 4 tablespoons fresh lemon juice, one-half cup of water. Process in a blender until pureed and serve immediately. Then sit back and enjoy a cup of 'Wild Blue.'"

"Our cattle are well fed" says Cathy. "With our rotational grazing system, they are moved to a new pasture with fresh grass every two days.
But walk in with a pail of salt and mineral and you can almost start a riot. They go animal over the mineral."

The First Lady

BRUCE MINES, ONTARIO

APPROXIMATELY 130,000 of the beef cattle in Canada — of a total of 3.7 million — are Black Angus. The breed, which originated in Scotland, was founded by Hugh Watson who began selecting and breeding the best black animals in his herd in the mid 1800's. His bull, *Old Jock* and cow *Old Granny* are documented in the Scotch Herd Book. We can trace the pedigrees of the majority of Angus cattle alive today back to these two animals.

In the small community of Bruce Mines, Ontario, Ron and Cathy Bonnett operate an 800-acre farm with 200 head of Angus. Ron is the president of the Ontario Federation of Agriculture (OFA) and his business travel takes him away from the farm 200 days every year. That puts the first lady in charge.

Cathy is a strong and confident woman who absolutely loves farming. "In the summertime, you should smell the sweetness of the clover and hay. It's so peaceful being out in the field. I think that's where I get my best therapy — in the early evening watching the animals at play.

"I'm lucky. Farming gives me lots of freedom and variety. I don't have to punch a time-clock. I love my work so much — my kids tease about having to pry me away from the farm." During haying and calving season, an average work day for Cathy is 10 to 12 hours long. At six each morning she checks on the bulls, cows and new calves, and then returns to the house to do bookkeeping and housework. After lunch, she's off to the fields to bale hay.

"I come from a want-to-be farming background. We pastured cattle and had a few pigs. My dad had a passion for it, but didn't want to fully take the risk and go all the way. I love being with the animals and I even enjoy field work. Baling is pretty straightforward. We have a round baler that makes 500 kg bales. Our tractor is equipped with a computer on the baler that gauges the size of bale and starts the wrapping process. It takes three minutes to bale, wrap and eject a bale destined for winter storage. On a really good day I can make about 100 bales. That's about a 35-acre-field."

Located just east of Sault Ste. Marie, Bruce Mines is a landscape of trees and rocks with Lake Huron at its doorstep. The township of just over 600 people is a tightly-knit community that suits Cathy to a tee. "We are an old-style rural area where everyone looks after each other. Recently, a neighbor had a serious accident and became quadriplegic. The people in our little town raised over $25,000 to help out. It's a great feeling to know you belong and that there are people around who you can rely on.

Ron is president of OFA (Ontario Federation of Agriculture) and is a strong advocate for farmers having a united voice. Some might argue his new membership drive is becoming too aggressive.

"The agriculture community is so cool. Our commonality with our farming friends is great: we're all at the same place in life and have known each other for years. Anytime I'm feeling a little down or out of sorts, I can always pick up the phone. Our friends are the kind of people who will have you laughing by the end of the conversation."

Cathy enjoys working with Angus and laughs at her own description of the perfect cow. "She's black and she's beautiful! She has a nice neat udder, deep chest, full butt and gentle eyes. Our adults weigh 545 kg, and the bulls are up around 1,000 kg. They are very curious animals who are extremely good mothers, known for their heartiness and easy calving.

"We sell our calves to a specialty market. Our breeding operation allows us to ship around eighty, 270 kg calves a year to a buyer in southern Ontario. They sell for about $3.10 per kilogram. We wean the calves in December, and then they are raised on hay and grain. We don't use growth hormones or antibiotics."

Cathy believes success is achieved by those who experiment and try new things. "We do things differently. For example, our pasture is divided into 25 smaller pastures of five acres each. We move the cows every two days to a new field. That way, we get about three times as many feeding days from our pasture than if we went the traditional route and cut the hay. The other advantages are that we don't have the input (fertilizer) costs and the cattle get a high quality feed. At the same time, they don't strip the pasture — and they fertilize their own field."

Ron's schedule includes a great deal of international travel but says when he leaves for a business trip, he doesn't give the farm a second thought. "I trust Cathy completely. She has a great ability — and the patience — for working with the animals."

In the early 1970's, when Ron worked as a steam fitter, he was attracted to the independence and the challenge farming offered. That's when he decided to become a risk taker. "Even though I had a family to support, I left my job and borrowed every penny I could to buy our farm. Working the farm in those early days was tough. In the 1980's interest rates went sky high. Cathy and I had lots of sleepless nights." When asked what kept him going, Ron smiles and says, "I was young, full of pride and didn't want to give up. And I guess I wanted to keep those 'I told you so's' at bay! You see, we farmers are stubborn. It's something that gets us through hard times."

As OFA president, Ron serves a membership of 38,000 farmers. His role is to ensure that politicians at all levels of government understand the concerns of Ontario farmers.

His insight into the challenges farmers face and mindset of the rural community allows him to speak with great clarity about issues both here in Canada and internationally. "Farming has changed drastically. The perception of 'Old MacDonald' isn't reality. If it were that way, consumers would be paying 20 times more for food. Things are not like they were 30 years ago. A lot of farming today is done by university grads who are forward-thinking and embrace technology."

'The First Lady' — Cathy Bonnett. "I can't imagine a better life than farming — even with the long hours. When I'm out baling, I can do about 35 acres a day and make up to 100 bales. It's a satisfying feeling when you can see what you've accomplished."

Cathy and Ron Bonnett

Canada produced a whopping 1.45 billion kgs of beef in 2004. Beef production contributed $6.4 billion to our economy with the major export recipients being the United States and Mexico. Globally, Canada ranks 12th in terms of its cattle inventories.

Ron says that over the next 20 years small farms will continue to become more niche targeted and farmers will become more focused on producing energy. "Bigger farms will produce larger volumes of product, but unfortunately, the mid-size farm may be a thing of the past. They will have to go big or target niche. All farms will look at the energy side of things. We are energy hogs, but the opportunity is there for farmers to go beyond being growers of food. We can grow fuel. We can farm to improve air quality and the environment. With biofuels, think of what we can accomplish for the environment and for economic activity in rural areas. The impact will be huge."

Ron is crystal clear on why consumers need to be aware of farming issues. "Consumers have to understand what's in it for them. Every time they make a purchase decision, they are deciding to support or not to support their local economy. We have quality assurance programs here in Canada to help ensure food safety. A farmer here cannot buy pesticides unless they are certified. You don't have these kinds of guarantees when you buy foods from other countries."

Cathy and Ron are living authentically in a community they fully support. As Ron says, "Farming is more than a business — it's about families and communities. Like everyone else, we have our troubles coping in this busy society. But I love the people. They are direct and don't colour code anything. You always know where you stand with a farmer."

Don's "canola spiral" is a simple but highly effective seed cleaner. As the grain begins its downward spiral, centrifugal force causes the round seeds such as canola or peas to jump over the rim of the inside circle of spirals, into the larger, outside ones. The flat-shaped, slower moving weed seeds and other foreign matter continue their fall down the inner circles before exiting into a separate container. In a world of complexity, a machine with no moving parts or electronic circuitry is a rarity.

Pedigree Seeds

HOLLAND, MANITOBA

DON ZEGHERS WAS FRESH OUT of university and only 23 years old when he came up with a great idea. He could solve a huge problem for local farmers and launch a successful business for himself at the same time. Don had noticed that farmers near Holland, Manitoba were forced to truck their grain 45 kilometres away to be processed. So he borrowed $10,000 and built a seed processing plant. He saved his neighbours time and money and earned himself 250 loyal customers. Today, Don still processes seeds for his clients, along with running a 2,000 acre farm.

Although Don lives in the homestead where his parents, Art and Alma, raised him and his nine siblings, farming *was not* his childhood dream. "I didn't like throwing hay bales or milking cows. I didn't know what I wanted to do, other than work for myself and be my own boss. I was just a teenager when my dad and mom were ready to retire. They held a family meeting and divided up their 600 acres of land between two brothers and myself. At the age of 19, I had my first crop."

Holland is in the heart of the prairies surrounded by farmland that sells for approximately $2,000 an acre. The region has

(Left to right): Krista, Don and Leah Zeghers

small pockets of rich, black soil and is a high moisture area. The land, along with fairly long growing days, supports a diverse range of crops. Don's fields produce 91,000 bushels of flax, oats, wheat, barley, pinto, soya, corn, canola, peas, winter wheat and rye seeds. His wholesale and retail businesses — selling to both companies and to other farmers — employ four full-time and three part-time employees. The majority of Don's time is focused on sales and marketing.

The genetic make-up of a plant is important to a seed grower like Don who produces high quality seeds with genetic purity. Farmers looking to get an edge on Mother Nature purchase a superior product like Don's, because the healthier the seed, the heartier the plant.

During harvesting, a consistent high level of quality control proves crucial for producing pedigree seeds. Each seed crop is cleaned and carefully graded by width, weight, length and density. It is critical to ensure that no other seeds are mixed into the process as the final product must be pure. "There is a lot that goes into growing seeds," Don says. "Pedigree seeds are certified, registered and regularly inspected to ensure that we produce a quality product with no impurities. The flax seeds we grow, for example, are shiny, reddish brown with nice consistent colour. They're the size of a wood tick and sticky when wet."

Don has come a long way from his $10,000 investment and first few acres of land. Although expansion has brought success,

"Farm life means long hours and hard work. But at the end of the day, I wouldn't have it any other way. I enjoy being my own boss."

*"Even though I farm 2,000 acres, the return margins on grain are so thin that some years it's hardly viable.
My seed cleaning operation is a straight business decision. It's an investment of over $1 million but it's a stable business
and adds value to my product. And with Krista working off the farm as an educator, we do okay."*

it has also come at a cost. Farmers like Don find themselves moving away from traditional farming and into the world of big business. "Overseeing an operation like this brings lots of challenges. I put huge investments into the ground and have over $1 million dollars tied up in machinery and in the processing plant. The costs keep going up — I spend about $50,000 each year just on insurance. It's strictly a business now, not like in my dad's time.

"When I was six years old and just learning to drive a bike, I'd try to drive through a narrow cattle gate without scraping my knuckles. I thought that *if* I made it through, I had passed a kind of test. Well, I guess I'm still testing myself. I'm still trying new things, but I've learned a lot from past experience. In those bad crop years, I just focus on the next season. I know I'll do better next year. Just like in golf — my *next* swing will be better."

Rensje, farm cat Fluffy, and daughter Trudy share the evening milking. "My husband Jason and I already own 11 Holsteins and are adding more beef cattle,"
says Trudy. "We'd like to make a future in dairy farming but with the high price of quota, I'm not sure if we can afford it."

Red and White

PETER AALBERS IMMIGRATED to Canada in 1980 and has strong feelings about what it means to be Canadian. "I love it here. Even if they gave me a free farm back in Holland, I wouldn't go back! This is a country with a future and lots of freedom. We have wide open spaces and lots of nature."

As a young couple, Peter and his wife Rensje, diligently saved every penny toward their big dream: someday, they would go to Canada. When they finally moved, they settled on a 168-acre farm near Thunder Bay, Ontario. With only 32 milking cows and some used farm equipment, they began a modest dairy operation. The following year, the couple welcomed their first child into the world.

But just two years into their new life, they were confronted with distressing news — Rensje was diagnosed with Lupus. She was only 27-years-old, and since neither she nor Peter had any extended family nearby, they looked to the community for help. With the assistance of supportive neighbours and friends, they made it through those difficult early years. Today, the Aalbers own and operate a successful dairy farm with one of Canada's largest herds of Red and Whites.

Holstein cows are familiar to most people as the typical, docile, white cows with amoeba shaped black patches. The Aalber's raise the lesser known breed of Holsteins known as Red and Whites — appropriately named for their colours.

*Rensje and Peter Aalbers.
"I love Canada," says Peter.
"Even if they gave me a free
farm back in Holland,
I'd never go back to farm."*

All Ontario farmers must register with one of the three accredited farm organizations. Registration identifies all farms by location and type of farming, funds the organizations, and qualifies members for potential government subsidy and assistance programs.

These cows tend to have less muscle tissue and thinner coats, compared to other cattle, because they put all of their energy into making milk. Consequently, they are outstanding milk producers.

Peter has 120 of the gentle, good-natured Red and White cows, and he knows each one by name. The heifers are kept in tie-stalls and are moved outside to pasture in nice weather. His mature cows weigh about 550 kg. Put in another perspective, a chicken weighs 2 kg, an African elephant about 7,000 kg.

The milking system on the Aalbers farm allows 16 cows to be milked at the same time in what's known as a double-eight herringbone parlour. Before the milk is shipped to a Thunder Bay dairy, samples are taken to ensure that each tank of milk meets strict government quality standards. The farm ships 16,800 litres of milk each week. If used to produce ice cream, that amount of milk would make approximately 22,000, two-scoop cones!

A number of fascinating stories claim to tell the tale of how ice cream was discovered, but one of the most interesting is that of *Charles the First.* Once upon a time, hundreds of years ago, Charles I of England hosted a sumptuous state banquet for his friends and family. The superb meal included many delicacies of the day, but the "coup de grace" was yet to come. After much preparation, the King's French chef had concocted a new dish. It was cold and resembled fresh-fallen snow. The guests were delighted, as was Charles, who asked the chef not to divulge the recipe. The king wanted the delicacy to be served only at the Royal table and he offered to pay the cook 500 pounds a year to keep it that way. The recipe remained a secret until 1649 when poor Charles was beheaded. And then was the secret of the frozen cream was a secret no more.

Peter can count at least five generations of farming in his family, and he remembers the hard work as a young boy. "By the time I was eight, I was milking four cows by hand every night. Farming is still a lot of work. Even today, I put in an average of 16 hours a day. Rensje and I start in the barn at six in the morning to check on the animals and prepare the milking equipment. Then I have lots of outside chores, maintenance, and field work to keep me busy during the day. The second milking starts at six in the evening. For me, the farm comes first and my personal life comes second. It's the only way we can seem to get ahead.

"We chose Canada because it gave us the opportunity to start from scratch. If I had to do it all over again, there's only one thing I would do differently. *I would have come sooner!*"

"I have good memories as a boy in Holland," reflects Peter, "Of having my own Red and White calf. When I came to Canada, I was told this wasn't a viable breed. Something in me wanted to prove these people wrong and now I have one of the largest Red and White Holstein herds in the whole country."

Once Started, Half Done

SOME DAYS, THE FIELD IS RIMMED by blue sky and you can look out into four-foot-high bright yellow plants. The mustard is so pungent to your nose, you just know that it is giving off flowers and making pods. It's a smell that you have never smelled before."

As aesthetically pleasing as a mustard crop in full bloom might be, Dave Weber says that in fact, it's his lentil fields that are really the star attraction.

"Unlike mustard, there isn't a smell but lentils grow into a beautiful, lush carpet of white flowers. It's our lentil crop that made the farm. You'll never go hungry when you farm lentils."

Lentils are most likely one of the oldest cultivated legumes. Each year, Dave's farm produces 35,000 bushels of the tiny, protein-packed pulses. "Pulse" refers to the edible seeds of lentils, peas, beans, and other plants with pods. Lentils are a rotation crop that replenishes the soil with nitrogen, reducing

Dave Weber: industrious farmer, astute businessman. "To survive today, you can't just be a farmer doing your field work. You've got to run a sharp business. I spend at least a third of my work week in the office with sales and marketing."

the need for chemical fertilizer. The round, flat, tiny bean comes in a variety of colours and is not only packed with protein, but with fiber, vitamin B1, minerals and tannins. Lentils are commonly used throughout the world as an inexpensive and nutritious source of protein, consumed in soups, or prepared with meat or rice.

Canada is the largest lentil exporter in the world and in 2006 exported 670,000 tonnes. The world trade in lentils is a multi-million dollar business, so it's no surprise that a good part of Dave's day is occupied with selling his product. "One third of my work week is spent on bookkeeping, sales and marketing. Sure, I'm out in the field on the tractor, but I am also on the phone finding a market and arranging "the deal." Canada's diet doesn't include a lot of lentils so much of my product is exported. Compared to the rest of the world, we really are non-users of pulses.

"I have good confidence with the people who market my crops. I've found out that the bigger you get, the more they want to deal with you. Buyers like volume. And it's all about becoming more efficient.

"I've seen farmers who have taken over their parents' land and really struggled. They haven't realized that things have to change—growth has to happen. Our farm was the first to grow lentils in our region, and we've made sure we kept up with technology. It's great! We have GPS, Blackberries, air conditioning in our tractors — I love it.

"My dad came over to Canada from Russia in 1911. Things have changed drastically since his time. He was an entrepreneur, and he had the first tractor and the first combine in the area. He had me helping with farm work when I was only eight years old.

Christmas was a fun time. I remember every year our family would get a case of Orange Crush and a box of mandarins. It seems like a small thing now, but back then it was special.

Dave and his wife, Cheryl, were high school sweethearts who recently celebrated 35 years of marriage. Cheryl fondly remembers the early days: "Dave and I met in grade eleven. Our very first date was at the movies to see Sydney Portier in To Sir With Love. We married five years after that and have raised three children here on the farm."

Dave says that although he loves the challenge of farming, his biggest ongoing hurdles are limited cash flow and anticipating a constantly changing consumer market. "We put tremendous investments into our crop for a short, 110 day growing season. And we have to be sure we grow the right crops to meet demand. Our family risks everything on the farm, but we are determined. We will keep working as long and as hard as it takes. My motto is 'Once started, half done!'"

Each spring Dave gears up for what he happily describes as his favourite season. "In April everything is new. I get energized and full of anticipation for what's ahead. It's

To drive back-country roads in parts of Saskatchewan is to see as many as four or five abandoned homesteads for every one occupied. The original settlers in 1880 1900 typically broke a quarter section (160 acres) of land. Today farms are usually 10 times that size — or even larger. With the consolidation of land and declining number of farm families, has also come the demise of many once vibrant rural communities.

the time of year that I get back out in the fresh air," he says. "I love the smell of the soil and the sense of accomplishment I get when my crops have been planted.

"I think urban people are caught off guard when they visit a farm for the first time. We had someone come by a few weeks ago, and he was amazed. He couldn't believe that in my backyard I have hundreds of acres planted in crops. I have high tech equipment, and I'm running a business as efficiently as anyone else in the business world. I guess people just don't know what farmers are all about."

Couple a child's inherent curiosity of the world with a farm environment, and the learning opportunities are immense. Chelsey and her father have just discovered monarch caterpillars in their barley field. The worms are no threat, fortunately, and are left to continue their foraging.

Dance of the Barley

STE. ANNE, MANITOBA

ON A PERFECT SUMMER AFTERNOON, when the light is soft and the wind is playful, Manitoba farmer Brian Dueck finds a moment of unexpected magic in his field. Thick, even rows of barley blow in the breeze as thousands of spiked grain stalks, coaxed by the gentle wind, perform a perfect prairie dance. Light shimmers and the expanse comes alive as smaller ripples of grass multiply into larger and larger waves of dark green. This field of visual poetry is just one of several that will contribute to the 40,000 bushels of barley Brian will harvest this fall.

Barley was discovered growing as a wild grass, thousands of years ago. Christopher Columbus may have brought barley to North America on his journey to the new world. This multi-purpose grain is used as animal feed for cattle, swine and poultry and is also a comfort food for humans. It's tasty and healthy: low in saturated fat, sodium, and cholesterol and high in manganese, fiber, thiamin and selenium.

Barley plants grow about one metre high: at maturity, the kernels grow hairy, spike-like heads. Used world-wide in breads, cereals, and soups, malting barley is also a key ingredient in beer

From the top of the 14-metre-high grain elevator leg, Brian has a birds-eye view of his farm. "It's good for checking for snow melt in the spring and, how much water is left in the fields after a heavy rain. And sometimes, later in the season when the crops are good, I'll go up just for the view. It's a satisfying feeling to know you've worked side by side with nature and have made it again. When it's not so good, I just stay off the elevator."

Work, play, and family go hand in hand on the family farm. Chelsey Dueck is oblivious to the giant sprayer looming above her mud-pie kitchen.

and whisky production. In 2005, it was grown in about 100 countries in the world, with the top five producers being Russia, Canada, Germany, France, and Ukraine.

Brian farms 1,200 acres of his own land and another 3,000 with his brothers. In addition to barley, they grow wheat, canola, oats and soya beans, and they raise hogs. "Our land is flat, heavy clay soil that gets really hard after a rain. The grain is used for seed production as well as for animal feed. In my area, I'm a bit of an oddball," says Brian. "Because 90 percent of my customers are farmers who buy the grain for seed."

Brian is the son of a grain farmer and he believes expanding and producing more volume is the only way farmers in the west can survive. "In my dad's time, he was able to raise us eight kids on a limited land base. I couldn't do that today. We have had to keep expanding our acreage and the number of hog barns — it was the only way we

The Duecks, (left to right): Brian, Chelsey, Kamera and Lydia

could continue. Things are always up and down for us, but we are fortunate to have enough livestock to keep things going when the crops fail.

"Our world grain reserves are less now than they've been in several years," says Brian. According to a 2006 report in *Agriweek*, a farm-business weekly, the world could face cereal shortages on a global level in two to three years. The report concluded that "The world is using more grain than it's growing; no reversal in sight…World cereal trade patterns for many years have shown relatively small changes in volume compared to changes in production, consumption and supply."

Brian was raised in the same farm house that he and his wife Lydia now share with their two children. He says the farm is a wonderful place for kids and the perfect location for an old-fashioned game of Cops and Robbers. "My brothers and I used a stack of 30 hay bales as a fort and chased each other around for hours shooting rubber bands. We had lots of fun!"

Lydia agrees that there's no better environment for her children than their own farm yard — or for her, for that matter. "Chelsey and Kamera can roam free here," she says. "They love to help me in the vegetable garden. We grow everything from strawberries to carrots, but peas are their favourite. There's nothing like snapping fresh garden peas right off of the vine. They're still warm from the sun when you pop them into your mouth!"

Lydia says that the farm can also be a wonderful place for children to learn life skills. "It's great to teach them about growing food. Last year, Kamera grew 26 pumpkins and sold them. She used the $11.25 she made to pay for our family's rides at the fair. At only seven years old, she's learned about sharing and the true value of money."

For Brian, the true reward of farming comes with the exhilaration of overcoming the odds. "If I didn't enjoy farming I wouldn't do it. It's true what they say — if your job is enjoyable, you are never at work. In the years that I have a good crop, there's no feeling like it. I know that this time, I've beaten the odds. I've worked with Mother Nature and I won! It's an adrenaline rush that you can't get anywhere else."

Of all the farm crops, few have a shorter harvesting window than hay. At least two days of good drying weather are essential to season fresh-cut hay before hauling it to the barn. Farmers pay close attention to radio and TV weather briefings, but most rely in large part on their own sense for predicting blue skies or summer thunder storms. On those rare perfect days, their balers and forage harvesters will be pressed into all-out action.

Cows and Quota

JACK RHYNER DONS GREEN OVERALLS, latex gloves and long, black boots, and begins his morning chores. It's a predictable routine. For Jack, milking cows is a 365 days-a-year job — but gone are the days when you pulled up a stool and *assumed the position*.

The parlour floor is sloppy with cow muck as eight well-behaved Holsteins saunter single file to the raised platform. Once they're in formation, Jack washes each cow's teats and udder with a red dye disinfectant solution. He then attaches vacuum cups to their udders and the milking machine extracts 14 litres of milk from each cow. The process is fast. Within four minutes, Jack removes the hoses. Immediately, the line-up of pleased-looking cows exits the barn. The next group gets into line and the cycle begins again.

Jack and his wife Debbie, have 150 Holsteins on a farm near Vermilion Bay in Northern Ontario. Jack says one of the reasons for increased milk production today, is the special care devoted to the animals' diet. "We dairy farmers get attached to our animals and hate to see them get sick. Technology has been a big help in that area. Our cows are chipped with a monitor around their neck which controls how much and what kind of feed they receive. That increases the overall health of our animals."

Cows have four stomachs — the rumen, reticulum, omasum, and abomasums — and they chew their cud six to eight hours a day. They consume a diet of 40 kg of hay, corn silage, dairy ration, mineral, salt, and they drink about 100 litres of water every day. Cows produce 90 percent of the world's milk;

(Left to right):
Bryan, Jack, Tianna
and Debbie Rhyner, with
dogs Wolf and Tido.

If there is one profession that teaches innovation and resilience, it certainly is farming. Out of necessity most farmers become pretty adept mechanics. "We operate several tractors, three trucks and at least eight different farm implements," states son Bryan. "Unless it's something specialized, we pretty well do all the maintenance and repairs ourselves."

just one healthy adult will give nearly 300,000 glasses in her lifetime. And each adult is an individual. A Holstein's spots are like snowflakes or fingerprints — every cow has a distinct pattern and no two cows' spots are alike.

Farmers in Ontario, Quebec, New Brunswick, Nova Scotia and Prince Edward Island belong to a regional pool known as the P5. The P5 share in all of the revenue for milk sales and in costs for transportation and promotion. Ontario has 4,800 dairy farmers and produces 2.5 billion litres of milk each year. The industry operates under a national supply management system which regulates imports and domestic production to ensure that production equals demand and also meets the average farmer's costs.

Milk prices at the retail level are not regulated in Ontario, but they are in some other provinces. Today, the price of a litre of milk is about $2.60. In 1958, the year that Jack Rhyner was born, a litre of milk cost about 20 cents. For every litre of milk purchased by the consumer, only a quarter of that purchase amount goes back to the farmer.

Dairy farmers must own a quota that controls the amount of milk that they can produce and sell. Milk quotas are bought through a quota exchange. The price fluctuates based on demand and on the amount of quota for sale. "The supply management system means we control how much milk we supply and that we are given a fair price for our product," says Jack. "One of the major problems we face today is the amount of imported butter oil and sugar blends that are coming across the border and being used as dairy substitutes. It means we are losing quota."

Today, the price of quota — which is effectively the privilege to sell the milk from a single cow — is $30,000! A cow might cost only $2,000. Jack says that very few young people today have the option to become a dairy farmer. "It is really difficult to get into the dairy business unless you get into it gradually or inherit it. With quota prices so high, it's virtually impossible for the next generation to get started.

"The pay-back formula works on the number of cows you can milk. Generally, 40 cows will give one person an income, and 80 cows will cover the wages of two people, and so on. But if I took the next step and expanded even further, we would no longer be a family farm," says Jack.

A farmer's financial portfolio is complex, as the life savings of several generations may be heavily invested in land, equipment and livestock. Jack says the challenge to maintain a dairy business and also support several independent family members is one that many farmers like himself struggle with. "Rising fuel costs, lack of cash flow, and long, long hours make farming a demanding and sometimes stressful way of life."

Although his work days are long and he's often faced with nearly overwhelming challenges, Jack isn't about to give up. "I've farmed all of my life so I don't know what else I would do. I guess I'm just too stubborn to get out. And I love working with my cows — they're gentle creatures. You know, somehow, in farming, there are always more good things than bad."

"I love feeding the calves," laughs four-year-old granddaughter Tianna. *"They're kind of silly and get all excited when I come into the barn with a bottle. Sometimes they suck on my fingers and that feels really weird."*

Near Kamloops, B.C. This horse is happy to stay out of Deb's viewfinder as long as he is rewarded with the scent and a lick of her coconut perfume.

Afterword

Vermilion, Alberta. Ken Farkash begins to question Deb's experience driving his team.

LIFE IS CURIOUS. We open, close, and look through doors all the time, rarely thinking about their significance. Some doors wait patiently for us to saunter through. They linger open, giving us lots of time to peek in, or peep safely from a distance as we carefully weigh the pros and cons of crossing the threshold. They are sensible, predictable doors that offer little surprise.

Others, like sliding doors, are different. Dare to walk through one of these and you'd better move quickly! They open fast, and close even faster. These are the doors that smack you on the backside while you're trying to squeeze through. They are edgy, full of unpredictability, and possess the power to change your life in a heartbeat.

Somewhere during the winter months of 2006, a sliding door presented itself. Carl Hiebert asked if I wanted to join him on another one of his projects. It was the *AgVenture* door, and I hurried on through.

On May 31, as Carl and I embarked on our cross-Canada journey, we were — as east coast folks say — *some nervous*. I felt like a big-eyed kid, bursting with anticipation, as visions of swimming holes, corncobs dripping butter, and wild blueberries danced in my head.

My job was to come back with 39 farm-family stories. With years of interviewing experience under my belt, I wasn't concerned about getting the stories. But the thought of learning to drive a 10-metre motorhome, *while* going through the Rockies, was a different story.

I faced my fear and bonded with *Bella* — the most courageous motorhome in world — during a white-knuckled drive through a miserable rainstorm in the Coquihalla Pass, B.C. Over the next 12,000 kilometres, Bella and I survived endless dirt roads, too-narrow city streets, and wide-open highways. I was lost so often that I posted a sign above me which read, *You are lost, get used to it!*

AgVenture presented us with a repetitive routine and a grueling pace. Wake at 5:30 a.m., drive for several hours, photograph and interview for another five hours, and then do it again. Day after day.

I met farm families who opened their homes and hearts. We would begin as strangers, and then, as they shared bits and pieces of their lives, we made that more intimate connection that leads to friendship. I discovered that mostly, farmers are just like the rest of us: living lives filled with joy, love, pain, and struggle, and doing the best they can.

Deb, although dressed in a red T-shirt, manages to hold her ground during a buffalo shoot at a ranch near Vermilion Alberta.

The Kilback hog farm. Whitewood, Saskatchewan. It could be debated who was having more fun…Deb or the pigs.

I learned that my spirited AgVenture warrior, Carl Hiebert, is without a doubt the most tenacious, determined person I've ever known. Every day, I watched Carl climb from his wheelchair onto *Ol' Red* in his armour of yellow raincoat, black rubber boots, goggles, and *smiling face*. Hours later, we'd meet again when Ol' Red clanged and banged her way to our rendezvous at the next farm. Again, I'd watch. But this time Carl would be wind-blown, splattered in mud and often shivering from cold. As he slowly pulled his hurt, stiff body down from the tractor, he never failed to give a wink, to laugh and deliver his battle cry: "If this trip was easy, someone else would have done it."

Through it all, what did I learn? Lots! I learned I was chameleon-like and well suited to the gypsy lifestyle. I discovered that Canada is the greatest place in the world to live. En-route, I spotted moose, coyotes, marmots, bears, foxes, skunks and wild turkeys. I took in breathtaking landscapes, skyscapes, and rainbows. I met cowboys and shepherds. I went eye-to-eye with a buffalo herd. I chased hay balers and traveled on potato harvesters. I hung out in milking parlours and pig pens and played in barns filled with one-day-old chicks … And I LOVED every minute of it.

The *AgVenture door* presented much more than I could have ever have imagined. Carl and I laughed, learned, loved and cried. And from all this came *Keepers of the Land*. Now, as we plan our future and new adventures together, I know life will present many more magical doors. So I wait … impatiently at times … for the next mysterious sliding door to open up. *Happy Sliding!*

Deb Cripps

Provincial Overviews

CANADIANS ARE FORTUNATE to have an agriculture base that is extremely diverse, thanks to our country's hugely varied climates, soils and farming families. From cranberry bogs in Vancouver, to the endless grain fields of Saskatchewan, to a strawberry U-pick farm in central Newfoundland, we are prolific producers of a huge variety of crops. The provincial summaries that follow will give you a better appreciation of our agricultural richness.

Alberta

Alberta is famous for beef and cowboys, but its 50,000 farms produce significant quantities of wheat, barley, pork, poultry, milk and fresh vegetables. They generate about $8 billion a year in revenues from 52 million acres of farmland — 24 million acres in crops and close to 22 million acres in pastures.

Many of the cattle ranches are in the foothills of the Canadian Rockies, but when it comes time to add the final pounds to ready cattle for market, they're taken into huge feedlots in the Lethbridge area. There are so many it's known as "feedlot alley".

That area is also home to intensive irrigation farming, much of it by descendants of Japanese pioneers. They grow a variety of vegetables, 58,000 acres of potatoes and 30,000 acres of sugar beets.

Central Alberta features family farms growing crops to feed hogs, dairy cattle and poultry. The province is also home to many Hutterite colonies where families live in communes and run multi-million-dollar farm businesses. The Hutterites are entrepreneurs, helping Canada to develop exciting export markets. For example, a group of them are in an alliance with Maple Leaf Pork, using genetic fingerprinting so shoppers in Japan can trace a pork chop back to a specific farm and litter of piglets. All of these pigs, which are fed special rations, are processed through a plant at Lethbridge that is dedicated to the Japanese market.

To the north is the Peace River prairie, featuring large grain fields and production of grass and alfalfa seeds marketed around the world. It was the last area of Canada to be opened to agriculture and now is a thriving, fully-developed industry featuring both production and processing.

Alberta has so much good farmland tended by so many excellent farmers that it depends on exports. It was, therefore, hit especially hard when Bovine Spongiform Encephalopathy (BSE, or mad cow's disease) turned up in 2004 and markets around the world banned Canadian beef and cattle. Alberta's farmers are so tenacious and persistent, however, that cattle remain the number one commodity, worth more than twice as much as wheat and three times canola.

One of the major challenges is keeping the younger generation on the farms, now that the province's vast oil and natural gas reserves have generated a huge demand for employees.

British Columbia

Beautiful British Columbia is famous for its towering Rocky mountains and Douglas Fir trees. Only three percent of its 95 million square hectares is farmed and half of that is really just wild pasture for grazing cattle. Yet British Columbia ranks first, second or third in Canada for the production of many foods, including dairy, poultry, apples, raspberries, cranberries, greenhouse tomatoes and peppers and cucumbers, flowers and furs.

There are dramatic differences in the types of farming practiced in the far-flung corners of British Columbia. It's an intensive, expensive agriculture close to Vancouver which has some of Canada's finest soils and one of the best farming climates. The lower Fraser Valley and the eastern part of Vancouver Island are home to Canada's largest dairy farms, its biggest raspberry (13,000 kg per year) harvest, cranberries (27,000 kg per year) and 7,700 kg of blueberries per year.

This area is also home to the province's fastest-growing agricultural sector — greenhouse production of tomatoes, peppers, cucumbers, flowers, bedding and nursery plants. Because land is scarce, and development pressures are intense, the British Columbia government was the first in Canada to declare an agriculture preserve in the early 1970s. As of 2001, there were 4.77 million hectares in the reserve, 617 million hectares in crops, 233 million in pasture and 1.2 million hectares of "unimproved" farmland.

Nestled between two mountain ranges, the Okanagan Valley is typical of intensive orchard and vineyard agriculture in the British Columbia interior. It's a near-desert dry area, but blessed with rivers and lakes that supply irrigation water. British Columbia has more than 700,000 hectares under irrigation, second only to Alberta. The farmers in the interior produce about 136,000 kg of apples, 13,000 kg of grapes, 8,000 kg of pears and 4,500 kg of cherries per year. They are world leaders in many ways, helped by a team of dedicated federal-government scientists at the Summerland Research Station. One of their recent successes is sweet cherries that are bigger and sweeter than anything else on the market.

Further to the north, are some of the biggest ranches in the world — the 500,000-hectare Douglas Lake Ranch and the 30,000-hectare Quilchena Cattle Co., both in the Nicola Valley. Still further north and east is the Peace River area, a vast prairie where the summer sun barely sets. There are huge fields growing wheat, barley, canola and forage seeds for the world.

Each of these areas of British Columbia agriculture is unlike the other, but they do have one thing in common — the tenacity and eternal optimism of about 10,000 farming families doing whatever it takes to survive in a tough business.

Manitoba

Manitoba is the gateway to the West and Winnipeg is the national home to many of the grain companies and organizations that serve global markets.

Wheat is, of course, the big crop worth $400 million to $500 million a year, but as with all of Canadian agriculture, Manitoba's farmers have been aggressively diversifying to keep their financial heads above water. Canola is now second at $350 million to $400 million a year and the hog business has been booming and is poised to keep on growing from $750 million to more than $1 billion a year.

Brandon is home to the nation's largest hog-packing plant and there are plans to build the second-largest one in Winnipeg. Tens of thousands of tiny piglets leave Manitoba farms to be grown up to market weight in Iowa. Chickens, eggs and dairy farms are all gaining because the cost of production is lower in Manitoba than most parts of Canada.

The province has 8.5 percent of Canada's farmers and 21,000 farms embracing 7.6 million hectares of which 4.7 million are growing crops. The rest is pasture or grazing land.

The richest farming area is in the Red River Valley, south of Winnipeg, and many of the farms there are cultivated by Mennonites who fled Russia in the early 1900s. Along with rich soils, southern Manitoba has the best climate of the province, allowing farmers to grow both traditional grain crops as well as a variety of other crops including peas, sunflowers, soybeans and corn.

New Brunswick

New Brunswick agriculture features meat and potatoes with emphasis on potatoes because Florenceville is home to McCain Foods Ltd., the world's largest manufacturer of French Fries.

About 23,000 hectares are planted to potatoes every year, producing a harvest worth about $100 million. Chickens and turkeys lead the meat sector with annual sales of about $57 million followed by hogs at $30 million and beef at $15 million.

New Brunswick is home to some of the world's best dairy cattle, even though most are sired by out-of-province bulls housed in Quebec. The province's dairy farms are among the largest and best run in the nation.

Cash receipts total about $420 million and are shared by about 3,000 farmers who own about 385,000 hectares, of which about 150 million are in crops. There are about 100 food-processing plants in the province.

The farming community is bilingual, and more balanced between French and English than any other province, including Quebec. Fredericton is home to one of the leading federal research stations and to the University of New Brunswick. The two co-operate to help the province's farmers and agri-food industry to develop and thrive.

Newfoundland

Newfoundland had the first farms in Canada, but the rocky island and northern climate make this a daunting place for farmers. Yet there are some, including more than a few successful commercial-scale operations, especially dairy and poultry farms.

There are more than 600 farms averaging about 18 hectares of cultivated fields each. Many of these, however, are small-scale part-time farmers. Livestock farming accounts for the lion's share of farm revenues at $65 million. Chickens bring in about $21 million and egg layers about $12 million. Crops account for about $17.5 million. Farmers have about 2,100 paid employees while food and beverage companies, which have sales of more than $400 million a year, employ about 1,700 workers.

Newfoundlanders, however, are famous for their tenacity, and few are more tenacious than the farmers who prove that this is a viable career and family choice.

Nova Scotia

Nova Scotia is famous for apples, but in recent years, milk has brought in more than 10 times as much money for the province's 3,500 farms. The Annapolis Valley, stretching 128 kilometres along the Northwest coast of the Bay of Fundy, is home to some of Canada's most picturesque farming communities featuring neat and tidy orchards, gardens and homesteads. The French planted the first apple orchards here in the 1600s and farming has been a mainstay of the region ever since, today featuring orchards, a wide variety of vegetables, livestock and poultry and related food-processing and farm-supply businesses.

The East coast, which lies exposed to the Atlantic ocean's fierce storms, has only small pockets of farming in the many coves and inlets. Each one is home to distinctive communities such as the tourist and artists' treasure of Penny's Cove and the colourful fishing town of Lunenburg.

There are 37,235 acres planted to blueberries, 350 beekeepers tending about 1,850 colonies and 113 fur farms producing about 765,000 pelts per year worth about $45 million. Not commonly known is a flower industry that brings in about $30 million a year.

Ontario

Ontario is home to Canada's largest, most diverse agriculture industry. From peas to pigeons, potatoes to pork, peppers to pears, it's produced and processed in Ontario.

Farm cash receipts routinely top $8 billion a year, close to $4 billion of that from crops and more than $4 billion from livestock, poultry and milk. Its richest counties produce more than any of the four Atlantic provinces, yet most farm parents lose their children to more lucrative careers in the nearby urban centres.

The supply-managed commodities — milk, eggs, chicken, turkey and hatching eggs — form a core of wealth counted in the billions for quota alone. That wealth is, however, under constant threat from increasing competition, especially world trade negotiations aimed at reducing tariffs. That makes politics a high priority, and lobbying a multi-million-dollar annual expenditure, for Ontario's farmers. Other hot political issues arise because of intense urban pressure and concerns about "factory farms", water pollution and food safety.

Yet the diversity of opportunities ensures that there is always some vibrancy in Ontario agriculture, such as huge investments in large greenhouses, in ginseng, organic farms and weaner pigs for export to the mid-Western United States.

While the farm population has declined as it has in every province, Ontario has a large number of small-scale farms that survive as part-time operations because the owners hold down full-time jobs.

The province features several distinct regions. Most of the farmers and production are in the Southwest. There are pockets of thriving agriculture around Thunder Bay and in the Clay Belt of the northern most farming area. The Niagara area features vineyards and peach and cherry orchards.

Prince Edward Island

Potatoes and red soil define agriculture on Prince Edward Island. About 105,000 acres are harvested every year and many end up in kitchens from Halifax to Toronto. Seed potatoes are marketed around the world. Two giant processing companies, McCain Foods Ltd. and Cavendish Farms, compete for spuds to make French Fries.

A constant challenge for potato growers are the wild price fluctuations. In 2005, when farmers across North America co-operated to voluntarily restrain production, P.E.I. revenues for potatoes were approximately $162 million. Other years, when revenues have been half that, farmers could only sell part of their crops.

Dairy farmers take in the second-most revenue — $62 million — with other crops and livestock trailing far behind. Hogs, for example, generate $27 million a year, cattle and calves about $18 million and eggs $4 million. Even though the island is small, there are about 1,800 farms embracing 646,000 acres.

The island's farmers and food processors have struggled. Chicken and turkey are gone because there is no processing company left to buy flocks and beef. Hog farmers, together with the provincial government, have had limited success with developing meat-packing facilities.

Quebec

As with all else, Quebec is a distinct society when it comes to agriculture and food. Its farms were laid out along the banks of the St. Lawrence River under the French seigneurial system and as sons matured, the farms were divided, each retaining river frontage. The result was hundreds of farms in narrow strips.

Dairy became a mainstay and Quebec developed a lock on more than a third of the Canadian market featuring integration from farms to co-operative processing plants. Through mergers, Agropur is now the largest dairy co-operative in Canada serving stores and restaurants across the country and some export markets. Through artificial insemination, some of the best dairy bulls in Quebec sire offspring around the world because Quebec dairy genetics are highly prized by leading dairy farmers.

In the mid-1960s, Canadian dairy policy began a shift to supply management, meaning farmers got full price for only a set volume of milk. That eventually evolved into a full-blown and compulsory supply management system run by producer-elected marketing boards, and while it guaranteed prosperity, it also curbed growth. Parents needed to look for alternatives to generate additional income to bring their sons and daughters into the business. Raising hogs seemed a viable alternative and within three decades, Quebec emerged as a leading province for hog and pork production.

Quebec agriculture developed around the church and the co-operatives and credit unions that priests helped develop. It also led to a highly-unified approach to farm politics and today, all farmers in the province must belong to the single farm organization to qualify for any government subsidies and services. The Union des Producteurs is the envy of farmers in other provinces for the influence it holds with provincial and federal politicians. And the co-operatives continue to dominate the meat, poultry, dairy and farm supply businesses in Quebec.

Quebec's farms are among the most modern and progressive in Canada. There are about 30,000 farms, 13 percent of the national total, and revenues run around $6 billion a year. The average farm is now approximately 30 percent larger than it was only 20 years ago.

Saskatchewan

Saskatchewan wraps its borders around 651,900 square kilometers, making it larger than many European countries. Within that area is more than half of Canada's cultivated farmland.

Much of it, however, is in an extremely arid area. Some farmers crop fields that are drier than the Sahara Desert. However, the rain that does fall usually comes precisely when it's needed to grow cereal crops, and explains why Saskatchewan grows the world's best bread-making wheat. The climate elevates protein levels, giving Canadian hard red spring wheats extra gluten strength to hold loaves together when yeast puffs up dough. The result is white, fluffy slices of delicious bread.

For decades, Saskatchewan exported shiploads of wheat to Europe. More recently millers and bakers have learned how to make the Canadian wheat stretch further, so Saskatchewan's farmers have sought out alternatives. Now, they produce durum wheat, world famous for pasta quality, canola, which is treasured for its heart-healthier vegetable oil, barley and oats. There are some surprises, such as $61 million worth of potatoes and $33 million worth of honey.

Crop revenues total about $3.3 billion a year, livestock revenues about $1.5 billion, led by beef with hogs and dairy a distant second and third.

One of the province's biggest challenges is maintaining rural communities. Larger tractors and combines, computers and satellite communications have enabled farmers to crop bigger farms, which has, in turn, depleted the rural population. It's difficult to maintain schools, churches and recreational facilities when there are so few people.

Making a Difference

INFORMATION AND EDUCATION are key tools to help us become more connected to our land and make responsible, informed decisions. The following list of farm facts and information provides just a glimpse of the complexity of issues facing farmers today.

Why should you eat locally produced foods?

THERE ARE SO MANY REASONS TO EAT LOCALLY, but at the top of the list is definitely taste and nutrition. Why eat a tomato that's travelled for days on a truck when you can enjoy locally grown plump beefsteak? That tomato is not only fresh picked and loaded with flavour, it contains more nutrients than one shipped from a distance, and it is also better for the environment. Today, the average food item travels more than 1,600 kilometres before it lands on our tables. Eating locally grown foods means burning less fossil fuel.

The significance of buying local can be startling. A study completed in the Waterloo Region of Ontario revealed that 58 commonly eaten foods (all of which could have been grown or raised locally) travelled an average of 4,497 kilometres from source to retail market. Substituting local foods for imports, for just this area alone, would result in reduced greenhouse gas emissions equivalent to taking 16,191 cars off our roads.

Five ways you can become involved and make a difference!

1. Buying Local
When you buy from local farmers, you are supporting local businesses and providing income for families in your own community. In a time when many farmers are struggling to maintain their livelihood, you can do your part to support them and preserve our rural heritage and keep our food dollars where they belong.

2. Pride of Canada
Request retailers do their part and buy Canadian. Does your lamb come from B.C. or New Zealand? Does your favourite restaurant offer a wide range of our great Canadian wines? Are the origins of products clearly identified? If not, demand that produce is clearly marked. Better yet, suggest that the manager dedicate a Pride of Canada area showcasing only Canadian products.

3. Education for your Family
Educate yourself and your children about food production and agriculture in Canada. Take the time to visit local fairs, farmers' markets and U-Pick farms to meet farmers face-to-face.

4. Share Your Concerns
Share your concerns about farming and rural issues with your local Member of Parliament. Let them know that farming in Canada is important to you.

5. Fundraising Opportunity
If you are interested in purchasing quantities of the *Keepers of the Land* book to help raise money for your club or organization, please contact the authors directly at: carl@giftofwings.ca or phone (519) 698-0051.

An excellent example of where to 'Buy local, buy fresh' — the St. Jacobs Farmers' Market, in St. Jacobs, Ontario. With more than 600 vendors, this has become the largest farm market in Canada, offering shoppers a year-round opportunity to deal directly with farmers and guarantee fair prices for both parties.

Did you know?

◆ Canada enjoys the third highest position in the world for the least amount of disposable income spent on food — approximately 12 percent.

◆ Just 100 years ago, over half of Canada's population were farmers. Today it's less than two percent. Just as remarkable is the increase in farming productivity: our great grandparents could produce enough food for 10 people. Today's farmer can feed more than 120.

◆ Average age of a Canadian farmer: 49

◆ Canadian farming is big business, with annual sales from crops and livestock production valued at more than $34 billion. As a nation, we are a net exporter by a wide margin with an annual trade surplus exceeding $5 billion.

◆ Farm families heavily subsidize their operations with off-farm employment contributing over 75 percent of total annual net income.

◆ Approximately 98 percent of all farms are still family-owned and operated although many involve more than one family member and operate as incorporated businesses.

◆ The amount of global cereal reserves (wheat, maize, rice and soya) peaked in 1986 to around 100 days. Today, our ability to feed the world with reserves is just over half that period.

◆ Ontario has the most number of farms — 59,728 and Newfoundland the least — 643.

◆ Farmers receive only a thin slice of the consumer food dollar. For a box of corn flakes which retails at $3.54, only 11 cents goes to the farmer for the corn. A beef rancher receives $1.83 for a prime sirloin steak that will cost you $14.04 in the store. Packaging sometimes costs more than what the farmer receives for the product inside the package. The gap between retail prices and return to the farmer continues to widen.

◆ Random testing by the Canadian Food Inspection Agency of over 40,000 food samples found compliance to be above 99 percent.

◆ While still small — less than 1.5 percent of all farms — organic food production in Canada is the fastest growing sector: over 20 percent annually.

◆ Farm products are used in the manufacture of an unlikely variety of items: corn, for example, is found in bicycle tires, toothpaste, degradable plastic, road de-icer and wallpaper.

◆ About 30 percent of Canada's agricultural land is not suitable to grow crops because it's too hilly, rocky, wet or cold — but it can support grazing livestock.

◆ Ethanol and bio-diesel hold great promise for future grain production. One bushel of soybeans produces about 1.5 gallons of biodiesel. In Ontario alone, implementing a five percent blend of ethanol in gas will create a market for 50 million bushels (approx. 300,000 acres) of corn annually and reduce greenhouse gas emissions by the equivalent of 200,000 cars.

◆ Corn accounts for 40 percent of the average Canadian's diet. It comes in the form of corn-fed beef, chicken, pork, milk, corn sweetener in soft drinks and processed food, flour and chips.

◆ A huge challenge faced by Canadian farmers is the unequal playing field created by U.S. farm subsidies — currently estimated at $21 billion a year.

◆ Farmers use less energy per acre today than 50 years ago, mainly because they have adopted low-tillage or no-tillage management. That has also significantly reduced soil erosion and water pollution.

Acknowledgments

An adventure of this scope, and subsequent book, happens only with many willing hands reaching out to help.
I am grateful to the following people and companies for their generous support:

Primary sponsors who made this venture possible in the first instance — McCormick, Cargill and Pioneer.

Triple E — for the generous loan of a luxurious motorhome.

Jim Rohman — for your continuous support from the very beginning.

Ernst Hofer — farmer friend whose creative mind and welding torch effected the necessary modifications for 'Ol' Red'.

Ag writers coast-to-coast who helped me identify our farmers en-route.

Riverbend Hutterites of Waldheim, Saskatchewan, and Noble Tractors in Kelowna, B.C. for speedy tractor repairs.

Casselman Express and Central Freightliners trucking companies for shipping the tractor to each end of the country.

Shaun MacLennen — master of website design and construction.

Glenn Fretz — logo designer second to none.

Nicole Battista and Katie Tyrrell, talented artists, for respectively supplying the art work and map in this book.

MKD/Child Financial Services for their support and Paul Born and Tamarack for being our on-the-road communications centre.

Gill Stead — for a first-class book design, and Kim Murouney for a keen editing eye.

Stoltz Farm Equipment — donation of an air-ride seat

Farmers — who generously gave us their time and shared their stories.

Elmira Lions Club — for having the vision to come on board to finance and distribute this book.

My life partner Deb — for doing the farm interviews and subsequent writing, the courage to wield a motorhome coast to coast, contributing creative photo ideas, and giving emotional support. And for simply being in my life. This is our book. We made it happen — together. Thank you Deb.

As one of the AgVenture sponsors, Cargill arranged an en-route celebration with employees and local farmers at its farm service centre in Morris, Manitoba.

We gratefully acknowledge the support
of these corporate sponsors who helped make
AgVenture a reality.

Central Cape Breton Island.